Springer Texts in Business and Economics

Springer Texts in Business and Economics (STBE) delivers high-quality instructional content for undergraduates and graduates in all areas of Business/Management Science and Economics. The series is comprised of self-contained books with a broad and comprehensive coverage that are suitable for class as well as for individual self-study. All texts are authored by established experts in their fields and offer a solid methodological background, often accompanied by problems and exercises.

R. K. Amit

Game Theory
with Applications
in Operations Management

R. K. Amit
Department of Management Studies
Indian Institute of Technology Madras
Chennai, India

ISSN 2192-4333   ISSN 2192-4341 (electronic)
Springer Texts in Business and Economics
ISBN 978-981-99-4832-1   ISBN 978-981-99-4833-8 (eBook)
https://doi.org/10.1007/978-981-99-4833-8

© The Editor(s) (if applicable) and The Author(s), under exclusive license to Springer Nature
Singapore Pte Ltd. 2024

This work is subject to copyright. All rights are solely and exclusively licensed by the Publisher, whether
the whole or part of the material is concerned, specifically the rights of translation, reprinting, reuse
of illustrations, recitation, broadcasting, reproduction on microfilms or in any other physical way, and
transmission or information storage and retrieval, electronic adaptation, computer software, or by similar
or dissimilar methodology now known or hereafter developed.
The use of general descriptive names, registered names, trademarks, service marks, etc. in this publication
does not imply, even in the absence of a specific statement, that such names are exempt from the relevant
protective laws and regulations and therefore free for general use.
The publisher, the authors and the editors are safe to assume that the advice and information in this book
are believed to be true and accurate at the date of publication. Neither the publisher nor the authors or
the editors give a warranty, expressed or implied, with respect to the material contained herein or for any
errors or omissions that may have been made. The publisher remains neutral with regard to jurisdictional
claims in published maps and institutional affiliations.

This Springer imprint is published by the registered company Springer Nature Singapore Pte Ltd.
The registered company address is: 152 Beach Road, #21-01/04 Gateway East, Singapore 189721,
Singapore

If disposing of this product, please recycle the paper.

Preface

Game theory is the study of competition and cooperation among rational, strategic, and self-interested agents. One of the application areas of game theory in recent years is operations management, which is the study of matching supply with demand under variability and uncertainty. Operations management is a core functional area in any organization in the manufacturing and service sectors. With advances in technology, operations management settings are increasingly becoming multi-agent, and game theory provides the right framework to study operations management in multi-agent settings. This is evident from the burgeoning literature employing game theory in operations management. Though there are excellent books on game theory, to the best of my knowledge, there is no book that connects game theory and operations management, and this is the first endeavor to provide an integrated study of game theory and operations management.

The objectives of this book are to provide an introduction to game theory and related solution concepts and to explicate these concepts with contemporary research problems in operations and supply chain management. The book has a modular structure. Chapters 1 and 2 provide an introduction to operations management and game theory, respectively. Traditionally, game theory is divided into strands—noncooperative games and cooperative games. Noncooperative games are represented as normal-form and extensive-form games, while cooperative games are commonly represented as characteristic-form games. Games in normal form and extensive form, along with the important solution concepts, are discussed in Chaps. 3 and 5, respectively. Games in characteristic form and related solution concepts are discussed in Chap. 7. Mechanism design and auctions are the high points among applications of game theory. Mechanism design deals with designing games by defining the rules, the action sets for the agents, and the appropriate equilibrium concept such that the designer's desired outcomes are the subset of the equilibrium outcomes of the game. Auctions are in the class of mechanism design problems. Chapter 9 provides a detailed discussion of the theory of mechanism design and auctions. Chapters 3, 5, 7, and 9 are complemented by Chaps. 4, 6, 8, and 10 on related applications in operations management, respectively. These chapters will assist a reader in understanding the game-theoretic skeletons that are the basis of complex operations management

v

models. The applications discussed in these chapters are based on seminal papers and highlight contemporary issues in operations and supply chain management. The applications include niche topics like explainable AI, blockchains, sustainability, physical internet, and community sensing. Each chapter has a problem set to check the understanding of the concepts.

How to Use This Book

The modular structure of the book facilitates adoption for multiple courses at different levels. The book can also be used for self-study. The draft versions of the book have been used for teaching an undergraduate course titled "Introduction to Game Theory" with 45 contact hours, and a graduate course titled "Operations Management in Multiagent Settings" for MBA and research students with 30 contact hours at IIT Madras. The following are the tentative recommendations for adopting the book for different courses:

1. A semester-long (45 contact hours) undergraduate course on game theory credited by students from multiple streams ranging from economics to engineering can cover Chaps. 2, 3, 5, 7, and 9 along with selected applications from Chaps. 4, 6, 8, and 10.
2. A quarter-long (30 contact hours) graduate course for MBA students can cover Chaps. 1, 2, 3, 4, 9, and 10 with emphasis on normal-form games and mechanism design. Alternatively, the course can cover Chaps. 1, 2, 3, 4, 7, and 8 with emphasis on normal-form games and characteristic-form games.
3. A quarter-long (30 contact hours) graduate course for research students with previous exposure to game theory can cover Chaps. 1, 2, 4, 6, 8, and 10.
4. A short course (15 contact hours) on mechanism design and auctions can cover Chaps. 2, 3, 9, and 10.
5. A short course for industry practitioners who use multi-agent architecture in business problems can cover selected applications from Chaps. 4, 6, 8, and 10.

Acknowledgements

My journey in game theory and operations management is influenced by my collaborators from academia and industry. In this regard, I deeply appreciate the support of Prof. Peeyush Mehta from IIM Calcutta, Prof. Kulwant Pawar from the University of Nottingham, Prof. Ashok Srinivasan from the University of Southern California, Dr. Shankar Venugopal from Mahindra and Mahindra, and Mr. S. Ramachandran from Infosys. I would like to thank the former and current research students, especially S. Dhandabani, Rajdeep Singh, and Prabhupad Bharadwaj, for numerous discussions and feedback on the topics discussed in the book. I also thank students of game theory and operations management courses at IIT Madras for providing valuable feedback that assisted me in presenting the material in an understandable manner. I am also indebted to the researchers and authors in game theory and operations management who have developed knowledge in these fields. I hope I have sufficiently acknowledged their contribution. I also appreciate the support of Ms. Nupoor Singh, Editor, Springer. Importantly, I would like to thank my wife Dr. Rashmi Gupta and my son Avik Gupta for their unwavering support during the preparation of this book. The completion of this book is a testimony to their patience. Finally, I thank my Parents for imbibing the spirit of excellence in me.

Enjoy the reading!

For any comments, I can be contacted at `rkamit@smail.iitm.ac.in` or `rkamit@iitm.ac.in`.

Chennai, India   R. K. Amit

Contents

1 Operations Management: A Curtain Raiser 1
  Problems 9
  References 9

2 Game Theory: Primitives and Representations 11
  2.1 Preferences and Utility 13
  2.2 Representations of Games 14
    2.2.1 Normal-form (or Strategic-form) 14
    2.2.2 Extensive-form 15
    2.2.3 Characteristic-form 17
  2.3 Game Theory Meets Operations Management 18
  Problems 19
  References 19

3 Games in Normal Form 21
  3.1 Examples 21
  3.2 Mixed Strategies in Normal-form Games 24
  3.3 Two-Person Zero-Sum Games 25
    3.3.1 Computing the Optimal Strategies for Two-Person
           Zero-Sum Games 28
  3.4 Solution Concepts for Games in Normal Form 29
    3.4.1 Pareto Optimality 30
    3.4.2 Domination 31
    3.4.3 Nash Equilibrium 33
    3.4.4 Preplay Communication and Correlated
           Equilibrium 44
    3.4.5 Bayesian-Nash Equilibrium 48
  Problems 52
  References 55

4	**Games in Normal Form: Applications in OM**	57
	4.1 Inventory Games	57
	4.2 Traffic Planning	61
	4.3 Airline Alliances	65
	4.4 Supply Chain Contracts	70
	4.4.1 Wholesale-Price Contract	73
	4.4.2 Buyback Contract	74
	4.4.3 Revenue-Sharing Contract	75
	4.5 Blockchains	77
	Problems	82
	References	83
5	**Games in Extensive Form**	85
	5.1 Examples	85
	5.2 Strategies in Extensive-Form Games	91
	5.3 Solution Concepts for Games in Extensive Form	95
	5.3.1 Subgame Perfect Nash Equilibrium (SPNE)	95
	5.3.2 Sequential Equilibrium	99
	Problems	103
	References	104
6	**Games in Extensive Form: Applications in OM**	105
	6.1 Capacity Decisions	105
	6.2 Leadership in Supply Chains	112
	6.3 Cheap Talk in Operations Management	115
	6.4 Extensive-Form Inventory Games	121
	Problems	123
	References	124
7	**Games in Characteristic Form**	125
	7.1 Examples	125
	7.2 Some Definitions for Characteristic-Form Games	128
	7.3 Solution Concepts	130
	7.3.1 The Core	130
	7.3.2 Shapley Value	133
	Problems	140
	References	140
8	**Games in Characteristic Form: Applications in OM**	143
	8.1 Inventory Centralization in Supply Chains	143
	8.2 Service Systems	148
	8.3 Towards Sustainability	150
	8.3.1 Recycling Coalitions	150
	8.3.2 GREEN Game	154
	8.4 Logistics Networks	155

Contents   xi

8.5   SHAP Algorithm 159
       8.5.1   Explanation Models 160
       8.5.2   Explanation Models in Quality Modeling 161
8.6   Aumann–Shapley Pricing 164
Problems 168
References 169

9   Mechanism Design and Auctions 171
9.1   Examples 171
       9.1.1   Procurement Problem 171
       9.1.2   Mechanism Design: Comments 179
       9.1.3   More Examples 181
9.2   Mechanism Design: Formalism 185
       9.2.1   Revelation Principle 187
       9.2.2   Procurement Problem Revisited 188
9.3   Auctions 193
       9.3.1   Desiderata for Auctions 193
       9.3.2   Efficient Auctions 194
       9.3.3   Optimal Auctions 199
       9.3.4   Combinatorial Auctions 205
Problems 208
References 209

10   Mechanism Design and Auctions: Applications in OM 211
10.1   Sponsored Search Markets 211
       10.1.1 Generalized Second-Price (GSP) Auction 213
       10.1.2 Vickrey-Clarke-Groves (VCG) Mechanism 217
10.2   Community Sensing 218
10.3   Physical Internet 222
10.4   Mechanism Design for Systems Engineering 227
Problems 229
References 231

Index 233

About the Author

R. K. Amit is Professor at the Department of Management Studies, Indian Institute of Technology (IIT) Madras, Chennai, India. He completed his undergraduate studies at Indian Institute of Technology (IIT) Kanpur and his doctoral studies at Indian Institute of Science (IISc), Bengaluru. He heads the Decision Engineering and Pricing (DEEP) Lab at IIT Madras, which specializes in engineering best decisions using wisdom from optimization, game theory, mechanism design, and pricing. His research has been published in top-tier journals. He is actively involved in high-impact, industry-relevant research.

Chapter 1
Operations Management: A Curtain Raiser

Around 1750, at the beginning of the industrial revolution, economic growth was negligible across the nations, and there was homogeneity in prosperity levels (Spence, 2012). With the introduction of steam and subsequent mechanization, productivity and per capita incomes started to rise in Europe, and the world witnessed "great divergence" in productivity. The revolution spread to the United States with the development of interchangeable machine-made parts and then mass production. The emergence of vertically integrated behemoths like Ford Motors and General Motors was the outcome of these innovations in production. The production paradigm was further changed with the introduction of "lean production systems" in Japan—Toyota Motor Corporation took the lead in evolving lean thinking. *Operations management* lies at the core of these production processes. This chapter discusses the core concepts in operations management and the related models.

One of the primitives of operations management is a *process* that is defined as any transformation from input to output.

Anupindi et al. (2013, p. 13) define operations as "business processes that design, produce, and deliver goods and services". To achieve high profitability, businesses strategize to deliver goods and services at low cost, with a rapid response, high variety, and high quality. The corresponding process competencies, as shown in Fig. 1.1, are *process cost, process cycle time, process flexibility*, and *process quality*.

Pound et al. (2014, p. 26) note, "Understanding the natural behavior of operations first requires a clear understanding of the environment every manager has to

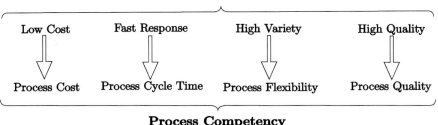

Fig. 1.1 Business strategies and process competencies

navigate". In operations settings, the environment is dynamically varying[1] and uncertain; and managers often have insufficient information about the environment. Information about the environment is another primitive of operations management. Additionally, managers improve process competencies, given the available technology. For example, production technology is such that the set-up cost is very high, which constrains a manager to achieve high process flexibility. In the real world, it is difficult for businesses to achieve all their objectives simultaneously due to insufficient information about the environment and technological constraints. Lower process cost can be achieved through higher utilization of available resources, resulting in higher process cycle time and slow response. Similarly, if the set-up cost is high, targeting higher process flexibility leads to higher process cycle time. In the traditional Ford production system with high set-up cost, Ford motors restricted the variety and produced "T-model" only in *black* color to achieve lower cycle time. Hence, we can define operations management as *improving process competencies under insufficient information about the environment, using the available technology*. Given the available technology, the set of possible actions or strategies to improve process competencies is another primitive. There are certain laws of nature,[2] like Little's law (discussed below), which relates actions and process competencies. These laws are also among the primitives of operations management. The above discussion is formalized in Definition 1.1.

Definition 1.1 If $\mathcal{P}$ is a process, A is a set of actions available to improve process competencies, $\mathcal{E}$ is the environment, then $\Pi_\mathcal{P} : A \times \mathcal{E} \to \mathbb{R}_+^n$. $\Pi_\mathcal{P}$ is n-dimensional vector associated with the process $\mathcal{P}$—each dimension indicates a *measurable* process competency. ◂

---

[1] Pound et al. (2014) define variability as the lack of uniformity. For example, in a manufacturing process, set-up times are not constant. For a service-oriented process, the arrival of customers or processing times are variable.

[2] Laws of nature are central in the evolution of science. In 1703, Issac Newton, as President of the Royal Society, defined science as "[g]eneral Rules or Laws,—establishing these rules by observations and experiments, and thence deducing the causes…" (Wootton, 2016).

1 Operations Management: A Curtain Raiser

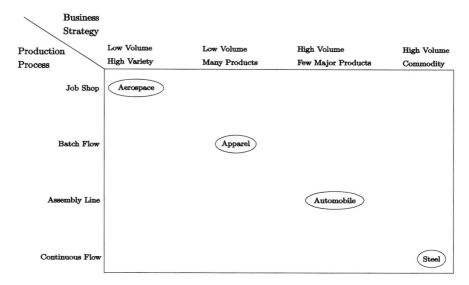

Fig. 1.2 Product-process matrix (Hayes & Wheelwright, 1979a)

In Definition 1.1, we assume that set A^3 is rich enough to include all possibilities, given the available technology, to a manager at the tactical, operational, and strategic level to manage the competencies of the process in the direction of business strategies.

A time-honored framework that matches a firm's business or competitive strategy with production processes is due to Hayes and Wheelwright (1979a, b) also known as *product-process matrix*, shown in Fig. 1.2. The product-process matrix provides reasonable guidance for the operations manager to synchronize the process with the business or competitive strategy of the firm. For instance, in order to differentiate on cost, a process ideally should maximize the capacity utilization of resources, leverage scale economies, simplify the design, and standardize the process in general. The process design for this low-cost strategy has to be consistent. A low-cost business model would need an inflexible, rigid process that facilitates a low-cost strategy. On the other hand, firms competing on high variety and customization, with all the complexities of demand uncertainty, would need a highly flexible process that allows

---

[3] In Definition 1.1, $A \times \mathcal{E} \to \mathbb{R}_+^n$ captures the essence of one of the most important *shloka* (or verse) in Indian philosophy, mentioned in *Bhagavad Gita* (Chap. 2: Verse 47)

> कर्मण्येवाधिकारस्ते मा फलेषु कदाचन ।
> मा कर्मफलहेतुर्भूर्मा ते सङ्गोऽस्त्वकर्मणि ॥४७॥
>
> karmaṇy evādhikāras te
> mā phaleṣu kadācana
> mā karma-phala-hetur bhūr
> mā te saṅgo 'stv akarmaṇi

which means that we can only control our actions, and not the outcomes.

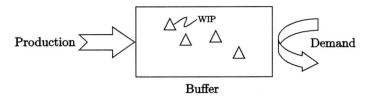

Fig. 1.3 Inventory buffer in a production process

buffer capacity to absorb demand fluctuations, allows buffer inventory to manage the demand service level, and allows frequent set-ups in the production process to serve variety with acceptable cycle time, and has, in general, a flexible and responsive supply chain to customize the product requirements. A job shop producing high-volume commodities leads to opportunity costs, while a continuous flow producing low volumes with high variety results in higher costs.

The product-process matrix highlights the importance of buffers. In hydrology, a control structure creates a reservoir (or a *buffer*) that synchronizes downstream demand with variable runoff. Similarly, a manager needs a buffer that matches demand with production in a production setting. Buffers also belong to the set of actions in Definition 1.1. Hopp and Spearman (2011) describe three kinds of buffers:

Inventory Work in process (WIP), or finished products waiting for demand—*parts or products waiting* (Fig. 1.3).
Time Delay in meeting the demand—*customers waiting*.
Capacity Extra resources that reduce the need for inventory and time buffers—*machine waiting*.

In service operations, only time and capacity buffers are available. As mentioned earlier, these buffers are the levers to achieve process competencies, and it is important to know the laws that relate buffers and process competencies.

One of the most important laws is *Little's law*, proved by Little (1961), which relates average work in process (I), throughput (R) (average number of units flowing through a process per unit time), and average cycle time (T) as

$$I = R \times T \tag{1.1}$$

To operations managers, Little's law provides insights into average inventory I requirements to achieve desired throughput R, for a given cycle time T. Hence, this provides a powerful tool to control the parameter of interest once any two parameters are known. This relationship, $I = R \times T$, captures the essence of performance measures in operations management.[4] The robustness of Little's law is due to its applicability in various conditions, independent of variability, as it considers the

---

[4] Originated from queuing systems, and the equivalent relationship is $L = \lambda \times W$ with analogous interpretation.

1 Operations Management: A Curtain Raiser

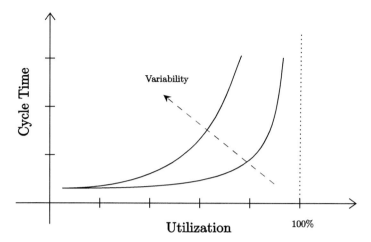

Fig. 1.4 Cycle time and utilization

average rate of the metrics—average inventory, throughput, and average cycle time. Furthermore, capacity is the upper bound of throughput.

Kingman (1961) establishes the relationship that average cycle time increases nonlinearly with capacity utilization. The relationship is called *Kingman's formula*. For given utilization, increasing variability of the environment increases cycle time (Fig. 1.4). As in Definition 1.1, given the environment, a manager can decide on the utilization level to achieve a target process cycle time. We refer Hopp and Spearman (2011) for a more detailed discussion on these issues.

There are other laws of nature that help in improving process competencies. One such law is *aggregation reduces variability*. Actions like "inventory centralization" (refer Sect. 8.1) or "capacity pooling" reduce the variability of the environment. Actions like outsourcing enable better technology and enhance capacity and process flexibility. However, outsourcing increases coordination costs—there is extensive literature on supply chain coordination that focuses on aligning incentives to improve process competencies. Disruptive technologies like *blockchains* and *platforms* facilitate coordination at lower coordination costs. "Aeroxchange" (https://www.aeroxchange.com) is one such platform for supply chain solutions for the aviation industry.

The above discussion highlights the importance of buffers in operations management. This book uses buffers as the common thread to link operations management and game theory. In the rest of the chapter, we will discuss important models related to buffers. Silver et al. (2016) is an excellent introduction to these models and beyond.

The first introduction of mathematics to operations management is due to Harris (1913).[5] The *Economic Order Quantity* (EOQ) model proposed in Harris (1913) has been serving the inventory problems for more than a century now and forms

---

[5] An interesting description of distortion of this citation in years to follow is in Erlenkotter (1989).

the basis for inventory policy in virtually every setting. The basic EOQ problem captures the fundamental trade-off in production settings—in a multi-product environment that requires frequent changeovers due to switching of products to meet the customer demand, the flow rate of the factory is reduced due to the time lost in product changeover (commonly known as set-up cost or time). Reducing the changeovers attracts a penalty for increased inventory cost for a given demand rate. The optimization problem is balancing the inventory and set-up cost penalties. The basic humble EOQ model has certain assumptions of deterministic demand rate, instantaneous production, and unit cost of the item not dependent on scale. Given demand rate R, unit inventory cost H, and set-up cost (S), the economic order quantity (or optimal order quantity that balances the inventory cost and set-up cost) is

$$Q^* = \sqrt{\frac{2 \times S \times R}{H}} \qquad (1.2)$$

The structural insights drawn from the EOQ model are compelling and lead to several significant results in inventory problems. Andriolo et al. (2014) and references therein provide a recent review of the EOQ model and its extensions over the past few decades.

In Eq. 1.2, the economic order quantity is proportional to the square root of the demand rate. It means that inventory centralization brings economies of scale. Another important situation arises when there are multiple items at a single replenishment point. It could be a production environment characterized by product variety and limited capacity to be shared for these items or a retailer's problem of stocking multiple items from a common supplier. These situations lead to coordination issues arising from the production or replenishment of multiple items. The benefits of coordinating for replenishment of multiple items would typically be in the form of reduced costs arising from economies of scale in production, transportation, and fixed ordering costs. However, these cost reductions would come at the expense of higher inventory and an overall reduction in flexibility at the level of the stock-keeping unit. The fundamental multi-item problem is determining the order quantity and frequency of each product to be ordered under coordinated replenishment. Goyal and Satir (1989) review the basic model under deterministic demand and stochastic demand settings.

In addition to the EOQ model and its extensions in inventory policies, another inventory problem that has formed the backbone of inventory theory is the *newsvendor problem*. In its basic form, the newsvendor problem captures the trade-off of under-investment and over-investment in inventory. The classical version of the newsvendor problem is for the context when demand is uncertain, and the demand is for a relatively short selling season, like style goods. The decision-maker has to commit to inventory decisions before realizing demand. At the end of the selling season, the outcome could be excess inventory (overage) or opportunity loss due to lost sales (underage). Based on the associated overage cost (C_o) and underage cost (C_u), the problem is to optimize the inventory decision. For a cumulative distribution function $\Phi(\cdot)$ of demand, the optimal inventory Q^* is

$$Q^* = \Phi^{-1}\left(\frac{C_u}{C_u + C_o}\right) \qquad (1.3)$$

The relationship in Eq. 1.3 is known as *critical fractile formula* and gives the *optimal service level* that is the probability of no stockout. The robustness of the newsvendor problem is due to the similarity of trade-offs in other critical operations problems, such as capacity management.

As mentioned earlier, aggregation reduces variability—aggregating n independent demands, each with the standard deviation σ, reduces the standard deviation to $\frac{\sigma}{\sqrt{n}}$. This is known as *square root law* (Anupindi et al., 2013). This law is the basis of some of the powerful ideas in operations management like *postponement* (or *delayed differentiation*)—postponing differentiation of products reduces the variability of demand and leads to better matching of demand with supply. This strategy is one of the toolkits to enhance process variety. For example, Essilor, the largest maker of spectacle lenses, uses this strategy in its two-stage production process. In the first stage, Essilor manufacturers "plastic blanks" in about 4,00,000 types; in the second stage, the blanks can be personalized into 120 million types of products in a matter of hours, closer to the customer (Marsh, 2013). Oeser and Romano (2016) study the impact of square root law in German manufacturing. In service systems, pooling capacity reduces variability and improves cycle time for the given capacity (Recall Fig. 1.4).

In the late 1990s, many firms realized that focusing on core competencies enhanced overall competitiveness, resulting in outsourcing many of the business activities for capacity and cost flexibility. This reduces levels of vertical integration in firms and the emergence of supply chains. These developments resulted in scenarios of conflict and cooperation among the supply chain partners. Hence, supply chain management was no longer a simple flow of products from suppliers to customers. It includes coordinating operations among the firms in a supply chain while creating overall value in the supply chain. One of the early references to the term *coordination theory* is in Whang (1995), which sets up a template for classifying issues based on coordination within operations, inter-organizational, and cross-functional coordination.

One of the key inhibitors of coordination is *double marginalization* in supply chains. Double marginalization arises due to the misalignment of incentives in supply chains, especially in sharing inventory risks. For example, in a retailer-manufacturer supply chain with uncertain demand, the retailer orders less inventory than the supply chain's optimal inventory to minimize inventory risk. Supply chain contracts are one of the coordination mechanisms to align the incentives. Cachon (2003) is an authoritative review of supply chain contracts. Using the newsvendor problem framework, some contracts studied for supply chain coordination are buyback contracts, revenue-sharing contracts, and quantity-discount contracts.

The other issue in supply chain coordination is information sharing. For example, small variability in a retailer's demand inflates to high variability as we move upstream in supply chains, as shown in Fig. 1.5. This phenomenon is known as the *bullwhip effect*. Being closer to the market, the retailer has better information about

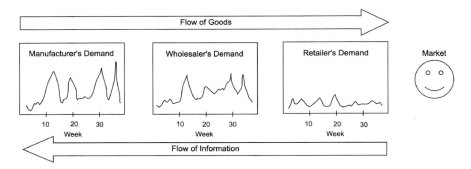

Fig. 1.5 Bullwhip effect

the demand, which gets distorted along the upstream supply chain. According to Definition 1.1, it means that there is asymmetric information about the environment $\mathcal{E}$ that hinders improving process competencies. Lee et al. (1997) identify factors such as demand signal processing, rationing, order batching, and manufacturer price variation that cause the bullwhip effect. Cachon et al. (2007) study the impact of the bullwhip effect across industries. Strategies like electronic data interchange (EDI) and vendor-managed inventory (VMI), which reduce information distortion, have helped overcome the bullwhip effect. The other possibility is to design contracts that align incentives for the exchange of truthful information with the upstream firms in supply chains. Cachon and Lariviere (2001) and Ren et al. (2010) are some of the papers contributing to this literature. Bray and Mendelson (2012) investigate the bullwhip effect in a sample of 4,689 public U.S. companies from 1974 to 2008. They report about two-thirds of firms have the bullwhip effect; however, the magnitude of the bullwhip effect had substantially reduced after 1995, compared to 1974–1994, indicating the success of some of these strategies.

The majority of traditional operations management models assume decision-making by a single firm. However, the above discussion points out that the crucial problems of operations management mimic multi-agent systems. For example, inventory centralization among individual firms can reduce demand variability or design supply chain relationships to improve process competencies. Game theory is the right tool for modeling operations management in multi-agent settings. This book is an introduction to game theory, attempting to link it with applications in operations management. We discuss the primitives and representation of games in the next chapter, with a concluding section to understand the merging of game theory with operations management.

Problems

Problem 1.1 As shown in the Figure below, a car assembly process at a plant takes 100 min to complete. The plant assembles 240 cars in a shift of 8 h. Compute cycle

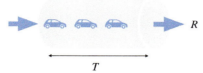

time (T), throughput (R), and average inventory (I) in the process. How can the plant manager reduce the average inventory requirement to maintain the throughput? ◄

Problem 1.2 In the product-process matrix (refer Fig. 1.2), identify the product-process mappings that are *economically unviable*. Also, identify the product-process mappings that are *technically infeasible*. ◄

References

Andriolo, A., Battini, D., Grubbström, R. W., Persona, A., & Sgarbossa, F. (2014). A century of evolution from Harris's basic lot size model: Survey and research agenda. *International Journal of Production Economics*.

Anupindi, R., Chopra, S., Deshmukh, S. D., Van Mieghem, J. A., & Zemel, E. (2013). *Managing business process flows*.

Bray, R. L., & Mendelson, H. (2012). Information transmission and the bullwhip effect: An empirical investigation. *Management Science, 58*(5), 860–875.

Cachon, G. P. (2003). Supply chain coordination with contracts. *Handbooks in operations research and management science* (Vol. 11, pp. 227–339).

Cachon, G. P., & Lariviere, M. A. (2001). Contracting to assure supply: How to share demand forecasts in a supply chain. *Management Science, 47*(5), 629–646.

Cachon, G. P., Randall, T., & Schmidt, G. M. (2007). In search of the bullwhip effect. *Manufacturing & Service Operations Management, 9*(4), 457–479.

Erlenkotter, D. (1989). Ford whitman harris and the economic order quantity model. *Operations Research, 38*(6), 937–946.

Goyal, S. K., & Satir, A. T. (1989). Joint replenishment inventory control: Deterministic and stochastic models.

Harris, F. (1913). How many parts to make at once. *Factory the Magazine of Management*.

Hayes, R. H., & Wheelwright, S. C. (1979a). Link manufacturing process and product life cycles. *Harvard Business Review*.

Hayes, R. H., & Wheelwright, S. C. (1979b). The dynamics of process-product life cycles. *Harvard Business Review*.

Hopp, W. J., & Spearman, M. L. (2011). *Factory physics* (3rd edn).

Kingman, J. F. C. (1961). The single server queue in heavy traffic. *Mathematical Proceedings of the Cambridge Philosophical Society, 57*(4), 902–904.

Lee, H. L., Padmanabhan, V., & Whang, S. (1997). Information distortion in a supply chain: The bullwhip effect. *Management Science, 43*(4), 546–558.

Little, J. D. C. (1961). A Proof for the Queuing Formula: $L = \lambda W$. *Operations Research*.

Marsh, P. (2013). *The new industrial revolution: consumers, globalization and the end of mass production*. Yale University Press.

Oeser, G., & Romano, P. (2016). An empirical examination of the assumptions of the Square Root Law for inventory centralisation and decentralisation. *International Journal of Production Research, 54*(8), 2298–2319.

Pound, E. S., Bell, J. H., & Spearman, M. L. (2014). *Factory physics for managers: How leaders improve performance in a post-Lean Six Sigma world*. McGraw-Hill.

Ren, Z. J., Cohen, M. A., Ho, T. H., & Terwiesch, C. (2010). Information sharing in a long-term supply chain relationship: The role of customer review strategy. *Operations Research, 58*(1), 81–93.

Silver, E. A., Pyke, D. F., & Thomas, D. J. (2016). *Inventory and production management in supply chains* (4th edn). CRC Press.

Spence, M. (2012). *The next convergence: The future of economic growth in a multispeed world*. Picador.

Whang, S. (1995). Coordination in operations: A taxonomy. *Journal of Operations Management*.

Wootton, D. (2016). *The invention of science: a new history of the scientific revolution*. Penguin Books.

Chapter 2
Game Theory: Primitives and Representations

From ancient times, human history has been dotted with numerous instances of conflict and cooperation. Why do humans conflict or cooperate? With the invention of science,[1] the focus of modern science shifts from observing natural and social phenomena to constructing theories to explain the observed phenomena. Game theory is one such theory, rooted in mathematics, to explain conflict and cooperation. According to Morgenstern (1968), the modern origins of game theory date back to Leibnitz, who emphasized the need to model conflict situations in 1710. In 1712, Waldegrave formulated an initial version of the minimax strategy. In 1881, Edgeworth showed the congruence between game theory and economics. One of the early and an important result is the *minimax theorem* by von Neumann (1928). The publication of "Theory of Games and Economic Behavior" by von Neumann and Morgenstern in 1944[2] provided a logical foundation of game theory and its application in economic and social sciences. Nash (1950) introduces the concept of *Nash equilibrium* that forms the cornerstone of applications of game theory in multiple disciplines, from economics to biology.

The primitives of game theory include *players* (or agents) who are rational, self-interested, and have free will. Each player is endowed with an action set. There are rules (or protocols) that enable interaction among the players. Rules include the protocols of communication among the players. The other primitive is information about the environment in which a game is embedded. These primitives of game theory capture the strategic situations of conflict or cooperation. Given the rules and the environment, the joint action of players leads to an outcome, and each player has

---

[1] Wootton (2016) is an excellent introduction to the philosophy of science.

[2] In 1994, to mark 50 years of publication of this book, the Nobel Prize Committee decided to award the Nobel Prize for Economic Sciences to game theory. John Nash, John Harsanyi, and Reinhard Selten shared the Nobel Prize in 1994. We discuss the contributions of these laureates in the subsequent chapters.

© The Author(s), under exclusive license to Springer Nature Singapore Pte Ltd. 2024
R. K. Amit, *Game Theory with Applications in Operations Management*, Springer Texts in Business and Economics, https://doi.org/10.1007/978-981-99-4833-8_2

a preference over the possible outcomes. Each player is strategic—chooses the action that leads to her preferred outcome, knowing that other players are doing likewise. Each player has partial control over the outcomes. In game theory, we study *solution concepts*—given the primitives, which outcomes are reasonable?

Traditionally, game theory is divided into two strands—*noncooperative games* and *cooperative games*.[3] In noncooperative games, the modeling unit is a single player—each player chooses noncooperatively. In noncooperative games, solution concepts are called *equilibrium outcomes*. Nash equilibrium and its refinement are examples of solution concepts for noncooperative games. Communication among the players is modeled in noncooperative games. In cooperative games, the modeling unit is a coalition of players—what can a coalition of players achieve? In a cooperative game, communication among the players is exogenous to the game. The examples of solution concepts for cooperative games are the *Shapley value*—axiomatic fair division of the payoff to a coalition; and the *core*—how to divide the payoff to a coalition so that it remains stable?

Noncooperative games are represented in *normal-form* (or *strategic-form*) and *extensive form*, discussed in Chaps. 3 and 5, respectively. The most common representation of cooperative games is *characteristic form*, discussed in Chap. 7. Let us consider an example of a noncooperative game.

Example 2.1 There are two players—1 and 2. Player 1 and Player 2 are endowed with action sets $\{T, B\}$ and $\{L, R\}$, respectively. Each player is self-interested and rational, and knows that the other player is also rational and self-interested. The combinations of their actions lead to outcomes, as shown in the matrix below (these games are called *bi-matrix games*). In each box of the matrix, the first number is the payoff to player 1, and the second number is the payoff to player 2. The payoff matrix is known to both players (a game of complete information). Each player has a preference over the outcomes; however, each player has partial control over the outcomes. The rules mandate that each player choose an action from the action set simultaneously, and there is no communication between the players. In this game setting, what are reasonable outcomes?

		Player 2	
		L	R
Player 1	T	10, 10	0, 11
	B	11, 0	3, 3

Player 1's preference for outcomes are $(11, 0) \succ_1 (10, 10) \succ_1 (3, 3) \succ_1 (0, 11)$. Similarly, player 2's preferences are $(0, 11) \succ_2 (10, 10) \succ_2 (3, 3) \succ_2 (11, 0)$. You can observe that the preferences are conflicting. The best outcome for player 1 is the worst outcome for player 2.

---

[3] This nomenclature is misleading, as cooperation is possible in noncooperative games like in repeated games). Cooperative games are also called *coalitional games*.

2.1 Preferences and Utility

If player 1 tries to achieve outcome (11, 0) by playing B, player 2 responds by playing R. If player 1 tries to achieve her second-best outcome (10, 10), player 2 again responds by playing R. Conjecturing player 2's responses, player 1 prefers to choose B. Similarly, conjecturing player 1's responses, player 2 prefers to choose R. Hence, if the actions of players decide the outcomes, the reasonable outcome is (3, 3). This outcome is self-enforcing, and is not the best outcome for any player. This outcome is also the Nash equilibrium[4] of this game. If the players can coordinate their choices, they can achieve (10, 10), which is a preferred outcome over (3, 3) for both players. ◁

Example 2.2 (*Golden Balls*) Golden Balls is a British daytime game show that was presented by Jasper Carrott (Source: Wikipedia). In one of the episodes, an interesting illustration of ideas is discussed in Example 2.1. The relevant parts of the episode can be watched at the following link

The reader should note that the host permits preplay communication before the play of the game. ◁

2.1 Preferences and Utility

In the preceding discussion, we mention that "each player is self-interested and rational". What is self-interest? Who is rational? In this section, we define these notions. Preferences over the set of outcomes are fundamental in defining these notions. Self-interest means that these preferences are derived from the player's attitude towards the outcomes. This does not preclude a possibility when a player prefers an outcome where other players benefit.

In game theory, players often decide under uncertainty. As in Example 2.1, player 1 makes her choice when she has uncertainty about the choice of player 2, and vice versa. Furthermore, in some cases, a player is uncertain about the environment (games of incomplete information); or, maybe, imperfectly aware of what had transpired earlier in the game (games of imperfect information). von Neumann and Morgenstern (1944) propose an axiomatic theory of decision-making under uncertainty. They represent the uncertainty in the form of a lottery—a list of outcomes and their associated probability. Each agent has a preference over the set of lotteries. von Neumann and Morgenstern (1944) assume that the preferences satisfy a set of axioms to represent

---
[4] Desiderata of Nash equilibrium is discussed in Chap. 3.

preferences in the form of a utility function, and the agents behave as if they are maximizing the expected value of the utility function. The "expected utility theory" is the workhorse of decision-making under uncertainty, and the axioms describe rational behavior. We are not listing the axioms (we refer Kreps, 1988 for an excellent introduction to this topic); however, the essence of the axioms is that a rational agent is intelligent enough to rank all the outcomes, forms expectations when deciding under uncertainty, and is consistent. This book assumes that the players are rational, as per the expected utility theory.

The axioms of von Neumann-Morgenstern expected utility theory have been tested for describing the observed behavior of the decision makers under uncertainty. Since Allais (1953), many papers have reported violations of the axioms. Various behavioral theories that are better equipped to describe the observed behavior, like the "prospect theory" (Kahneman and Tversky, 1979) and the "rank-dependent utility theory" Quiggin (1993) have been proposed in the literature. Gilboa (2009) provides a good introduction to the theories of decision-making under uncertainty, and Camerer (2003) integrates behavioral theories and game theory.

2.2 Representations of Games

In this section, we study different representations of noncooperative and cooperative games.

2.2.1 Normal-form (or Strategic-form)

Normal-form (or strategic-form) game is a representation of noncooperative games. In this representation, each *strategic* player chooses her plan of action independently. Communication among the players is modeled and can vary from no communication to preplay communication. Preplay communication also called *cheap talk* by players to influence the outcome. In Example 2.1 with preplay communication, it is not costly for player 1 to lie that she will choose T. The outcome is uncertain as it depends on the players' joint actions.

Definition 2.1 (*Normal-form game*) A normal-form game is a tuple $\langle N, (A_i), (u_i) \rangle$. N is a finite set of n players, indexed by i. A_i is an action set of player i. The combination of actions chosen by each player gives the set of action profiles $A = A_1 \times \cdots \times A_n$. Each action profile gives a unique outcome. Each player has a utility function $u_i : A \to \mathbb{R}$, in the spirit of von Neumann-Morgenstern theory, which captures the preferences of player i over the set of outcomes. ◀

In this book, we use this representation to model simultaneous-move games. Example 2.1 is a normal-form game. Chapter 3 discusses normal-form games, with applications of normal-form games in operations management. This representation is

2.2 Representations of Games

generic enough to model sequential-move games using "induced normal-form"; however, extensive-form games are a more efficient way to model games with sequential moves.

This representation can also model strategic situations when the players have incomplete information about the environment—"games of incomplete information" that are usually modeled as *Bayesian games*, discussed in Sect. 3.4.5. Simultaneous auction, like *first-price sealed-bid auction*, is an example of a game of incomplete information where each player knows her valuation for the auctioned object but is unaware of the valuations of the other players.

2.2.2 Extensive-form

Extensive-form game is another representation of noncooperative games that can model sequential moves of players in a game, as a game tree.

Example 2.3 Assume in Example 2.1, if the rules are amended such that player 1 chooses from the action set $\{T, B\}$, then player 2 makes the choice from the action set $\{L, R\}$, and the choice of player 1 is known to player 2. This strategic situation can be shown as an extensive-form game in the form of a game tree (Fig. 2.1).

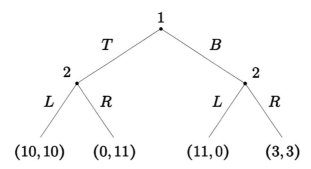

Fig. 2.1 The game tree

Definition 2.2 (*Extensive-form game*) An extensive-form game has N—a finite set of n players, indexed by i, and a game tree with the following structure:

- It has a set of nodes $x \in X$ and a binary relation $\prec$ called *precedence*. $x \prec x'$ means x precedes x' in a game tree. Each node in X has only one predecessor. $\prec$ is transitive ($x \prec x', x' \prec x'' \Rightarrow x \prec x''$), asymmetric ($x \prec x' \Rightarrow x' \not\prec x$), and incomplete (not every pair of nodes x, y can be ordered).
- The set of nodes with no predecessors are called *initial nodes*. The set of nodes that do not precede other nodes is called *terminal nodes*, denoted as $Z \subset X$.

- Every node $x \in X \setminus Z$ is assigned to a player $i \in N$, or to nature.[5] $A_i(x)$ is the action set of player i, if node x is assigned to player i. A probability distribution is known for the moves of nature.
- Each player has a utility function $u_i : Z \to \mathbb{R}$. $u_i(z)$ is the payoff to player i when the game terminates at $z \in Z$. ◂

In an extensive-form game with *perfect information*, a player knows her position perfectly in the game tree. In an extensive-form game with *imperfect information*, a player is imperfectly aware of her position in the game tree. In Example 2.3, if the choice of player 1 is not revealed to player 2; then player 2, when making his choice, is not sure whether he is on the left node or the right mode in the game tree in Fig. 2.1. *Information sets* capture a player's knowledge about her position in the game tree.

Definition 2.3 (*Information Sets*) The decision nodes in X are partitioned into information sets such that the following conditions are satisfied for any x and x' in the same information set:

- $x \not\prec x$ and $x' \not\prec x$.
- The same player i is assigned to x and x'.
- $A_i(x) = A_i(x')$. ◂

It means that the player cannot distinguish among the nodes in her information set. For a game of perfect information, each information set is a singleton. In a game tree, the nodes in the same information set are joined by a dashed line. The modified game tree of Example 2.3, when player 2 cannot distinguish between nodes x and x' in the same information set, is shown in Fig. 2.2.

In extensive-form games, the players are intelligent enough to make an assessment of what had transpired earlier in the game tree; and, using the assessment and conjecturing what will transpire at the subsequent decision nodes, choose the best action at each information set. We use extensive-form representation to model strategic situations with sequential moves in Chap. 5.

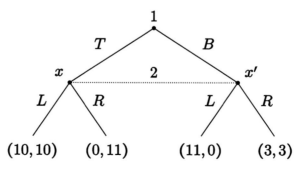

Fig. 2.2 Game of imperfect information

---

[5] In game theory, *nature* is a non-strategic player that captures exogenous uncertainty—uncertainty that does not depend on the choices of strategic players..

2.2 Representations of Games

2.2.3 Characteristic-form

In cooperative games, the modeling unit is a coalition of self-interested, rational, and strategic players. In cooperative games, communication among players is not explicitly modeled, and it is assumed that the players have explored all possibilities of communication and devised binding commitments to coordinate their actions. Given this assumption, the focus of cooperative games is on modeling payoff to coalitions using a *characteristic function*. In this book, we assume that the payoff to a coalition is transferable using a common currency—such class of cooperative games is called *characteristic-form games* (or coalitional games with transferable utility (TU games[6]).

Definition 2.4 (*Characteristic-form Game*) A characteristic-form game is a tuple $\langle N, v \rangle$, where

- N is a finite set of n players, indexed by i.
- $v : 2^n \mapsto \mathbb{R}$ is a characteristic function that assigns a payoff to each coalition in the game.
- $v(\emptyset) = 0$. ◄

$v(S)$ is a payoff to coalition $S \subseteq N$. $v(S)$ can be defined as the maximum payoff that the coalition S can achieve by joint action of the players in S, independent of the actions of the players in $N \setminus S$.[7]

Example 2.4 In Example 2.1, the rules are modified such that the players can communicate with each other and can sign binding agreements for joint actions. The maximum payoff that the players can jointly achieve is 20, when player 1 chooses T, and player 2 chooses L. The maximum payoff that each player can achieve, independently of the other player, is 3. Hence, $v(\{1\}) = v(\{2\}) = 3$, and $v(\{1, 2\}) = 20$. Any agreement that offers a payoff of less than 3 to either player will be rejected by the player. ◁

Definition 2.5 (Superadditive Game) A characteristic-form game $\langle N, v \rangle$ is superadditive if $\forall S, T \subset N$ and $S \cap T = \emptyset$, the following condition holds

$$v(S \cup T) \geq v(S) + v(T)$$ ◄

Superadditive games capture the situations when joint actions of coalitions S and T yield a payoff that is as large as the sum of payoffs S and T can achieve by acting independently. The characteristic-form game in Example 2.4 is a superadditive game

---

[6] We recommend Maschler et al. (2013) for a detailed discussion on cooperative games with non-transferable utility (or NTU games).

[7] A richer representation to model the influence of $N \setminus S$ on the payoff to S is *partition function form* (Tripathi and Amit, 2016).

$(v(\{1, 2\}) > v(\{1\}) + v(\{2\}))$. In this book, we focus on superadditive characteristic-form games. For superadditive games, coalition N known as *grand coalition* achieves the maximum payoff, and is collectively optimal for all the players.[8] For such a class of games, we study solution concepts, like the core that ensures the stability of the grand coalition; or the Shapley value that prescribes a fair division of $v(\{N\})$ among its members.

We discuss characteristic-form games to model strategic situations with coalitions, like cooperation in alliances, in Chap. 7.

2.3 Game Theory Meets Operations Management

We provided a curtain raiser on operations management in Chap. 1. Definition 1.1 represents operations management in a mathematical framework, with A as a set of actions available to improve m-dimensional process competency vector. Process competency outcome $\Pi_{\mathcal{P}}$ depends on set A and environment $\mathcal{E}$. Mathematically, it can be represented as $\Pi_{\mathcal{P}} : A \times \mathcal{E} \to \mathbb{R}_+^m$.

The framework can be extended to settings with multiple players (or agents). Definition 2.6 provides a mathematical framework of operations management in game theory settings. In Definition 2.6, agents (or managers) are rational, self-interested, and strategic. Process competency outcome $\Pi_{\mathcal{P}}^i$ of agent i depends on A_i, environment $\mathcal{E}$, and the choice of actions by the other agents.

Definition 2.6 If $\mathcal{P}$ is a process, A_i is a set of actions available to improve process competencies of agent (or manager) $i \in N$, $\mathcal{E}$ is the environment, then $\Pi_{\mathcal{P}}^i : A_1 \times \dots A_i \times \dots A_n \times \mathcal{E} \to \mathbb{R}_+^m$. $\Pi_{\mathcal{P}}^i$ is m-dimensional vector associated with process $\mathcal{P}$ for agent i—each dimension indicates a *measurable* process competency. ◀

In Chap. 1, we mentioned that set A is rich enough to include all possibilities, given the available technology, to a tactical, operational, and strategic manager to manage the process's competencies in the direction of business strategies. We now discuss the nature of the actions using the VCAP framework proposed by Van Mieghem and Allon (2015). VCAP stands for $V = C \cdot (A + P)$,[9] where V is the value that a manager wants to maximize using the right assets (A) and processes (P) that are aligned with capabilities (C). The components of capabilities, assets, and processes are shown in Fig. 2.3.

Action set A_i of agent i is on the asset or process side in the VCAP framework, and we indicate it for each OM application in Chaps. 4, 6, 8, and 10.

---

[8] It is possible to model situations when coalition structures, other than the grand coalition, are optimal. Maschler et al. (2013) is a good reference for characteristic-form games with coalition structures.

[9] We use *scalar product* to show that the value is only created when the assets and/or the processes are aligned with the capabilities.

References

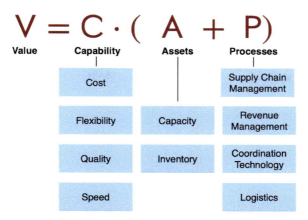

Fig. 2.3 VCAP Framework (Adapted from Van Mieghem and Allon 2015)

Problems

Problem 2.1 Consider Golden Balls game (Example 2.2). Show normal-form representation of the game when the players choose simultaneously. Compute the reasonable outcomes. ◂

Problem 2.2 Consider Golden Balls game (Example 2.2). Show extensive-form representation of the game when one of the players chooses before the other player. ◂

Problem 2.3 Consider Golden Balls game (Example 2.2). Show characteristic-form representation of the game when players can form a coalition. ◂

References

Allais, M. (1953). Le Comportement de l'Homme Rationnel devant le Risque: Critique des Postulats et Axiomes de l'Ecole Americaine. *Econometrica, 21*(4), 503.
Camerer, C. (2003). *Behavioral game theory: Experiments in strategic interaction*. Russell Sage Foundation.
Gilboa, I. (2009). *Theory of decision making under uncertainty*. New York: Cambridge University Press.
Kahneman, D., & Tversky, A. (1979). Prospect theory: An analysis of decision under risk. *Econometrica, 47*(2), 263.
Kreps, D. M. (1988). *Notes on the theory of choice*. Westview Press.
Maschler, M., Solan, E., & Zamir, S. (2013). *Game theory*. Cambridge University Press.
Morgenstern, O. (1968). Game theory: Theoretical aspects. In D. L. Sills (Ed.), *International encyclopedia of the social sciences*. Macmillan.
Nash, J. F. (1950). Equilibrium points in N-person games. *Proceedings of the National Academy of Sciences of the United States of America, 36*(1), 48.

Quiggin, J. (1993). *Generalized expected utility theory: The rank-dependent expected utility model.* Amsterdam: Kluwer-Nijhoff.

Tripathi, R. R., & Amit, R. (2016). Equivalence nucleolus for coalitional games with externalities. *Operations Research Letters, 44*(2), 219–224.

Van Mieghem, J. A., & Allon, G. (2015). *Operations strategy.* Belmont, MA: Dynamic Ideas.

von Neumann, J. (1928). Zur Theorie der Gesellschaftsspiele. *Mathematische Annalen, 100*(1), 295–320.

von Neumann, J., & Morgenstern, O. (1944). *Theory of games and economic behavior,* 1st edn. Princeton University Press.

Wootton, D. (2016). *The invention of science: A new history of the scientific revolution.* Penguin Books.

Chapter 3
Games in Normal Form

Chapters 1 and 2 provided the background for the remainder of this book. In this chapter, we consider normal-form games, a representation of noncooperative games, to model strategic situations when the players move simultaneously. We study different solution concepts for such a class of games and discuss their existence and computation. We begin this chapter with some examples.

3.1 Examples

Example 3.1 (*Prisoners' Dilemma*) This is one of the most commonly used examples to illustrate basic ideas of game theory. In the prisoner's dilemma, two prisoners are being convicted for a crime. Lacking credible evidence, the authorities question the prisoners independently to know whether they committed the crime. No preplay communication is allowed. Each prisoner has two possible actions: $\{C, D\}$—confess the crime (C) and do not confess (D). This is also a bimatrix game, and the payoff matrix is shown in Fig. 3.1.

Prisoner 2

		D	C
Prisoner 1	D	$-1, -1$	$-20, 0$
	C	$0, -20$	$-5, -5$

Fig. 3.1 Prisoners' dilemma

© The Author(s), under exclusive license to Springer Nature Singapore Pte Ltd. 2024
R. K. Amit, *Game Theory with Applications in Operations Management*, Springer Texts in Business and Economics, https://doi.org/10.1007/978-981-99-4833-8_3

C is the *dominant action* for each prisoner, irrespective of the choice made by the other prisoner. Assuming the prisoners are rational, self-interested, and strategic, the reasonable outcome is (C, C).

It is to be noted that C is the *best response* (see Definition 3.8) of each player, independent of the choice of the other player. The intersection of best responses is Nash equilibrium (see Definition 3.9). If the best response of Prisoner 1 is represented as a *rectangle*, and the best response of Prisoner 2 is represented as an *oval*, then (C, C) is also Nash equilibrium as it is the interaction of the best responses of each player. This is shown in Fig. 3.2.

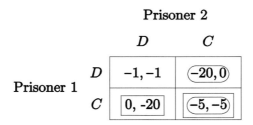

Fig. 3.2 Nash equilibrium in Prisoners' dilemma

Example 3.2 (*Matching pennies*) In the matching pennies game, players simultaneously choose between Head (H) and Tail (T). If the choices match, Player 1 transfers ₹1 to Player 2; if they differ, Player 2 transfers ₹1 to Player 1. The payoff matrix is shown in Fig. 3.3. This is an example of a *two-person zero-sum game* (or a *strictly competitive game*).

		Player 2	
		H	T
Player 1	H	$-1, 1$	$1, -1$
	T	$1, -1$	$-1, 1$

Fig. 3.3 Matching Pennies

It can be easily checked that the game has no outcome among the four possible outcomes, which is reasonable and self-enforcing. Using the rectangle-box method, there is no outcome where the best responses intersect. Is there an outcome beyond these outcomes that is reasonable and self-enforcing? We extend the concept of actions to *mixed strategies* to search for such an outcome.

3.1 Examples

Example 3.3 (*Cournot duopoly*) One of the key application areas of game theory is *industrial organization*—the study of imperfect competition among firms. Models in industrial organization are differentiated by the conjectures that each profit-maximizing firm makes about the actions of the competing firms. One of the classical models in industrial organization is the Cournot duopoly, in which two firms are competing simultaneously, and each firm assumes that the other firm will compete by choosing a fixed quantity of a substitutable good (Cournot conjecture). In this model, quantity is akin to capacity that, once decided, can be inflexible. Using Definition 2.1 of normal-form representation, $\langle N, (A_i), (u_i) \rangle$ with $N = \{1, 2\}$, $A_i = [0, 10]$, and $u_i = (10 - a_1 - a_2) \times a_i - c \times a_i$, is a Cournot game. Each firm i chooses quantity $a_i \in A_i$ to maximize its profit u_i, which is also dependent on the choice of the other firm. The resulting price is $(10 - a_1 - a_2)$, and the unit cost of production is c. It is worth noting that the action sets in this game, unlike in Examples 3.1 and 3.2, are not finite.

Firm 1, not observing the choice of firm 2 and using Cournot conjecture, can compute its reaction function (or the best response function)—optimal a_1^* as a function of a_2, which is $a_1^*(a_2) = \frac{10-c-a_2}{2}$. Likewise, $a_2^*(a_1) = \frac{10-c-a_1}{2}$. Using Cournot conjecture, a reasonable and self-enforcing outcome is an *equilibrium* (a_1^*, a_2^*) when each firm does not want to change the chosen quantity, knowing that the other firm also does not want to change the chosen quantity. The equilibrium can be computed as the intersection of the reaction functions, which is $a_1^* = a_2^* = \frac{10-c}{3}$. ◁

Example 3.4 (*First-Price Sealed-Bid Auction*) Game theory is the language of modern auction theory. A first-price sealed-bid (FPSB) auction is one of the classical auctions where bidders bid simultaneously in sealed bids for an item. The bidder with the highest bid wins the auction and pays her own bid to get the item.

Let us consider a case with two bidders who are privately informed of their value for an item; however, each bidder knows that the value of the other bidder is uniformly distributed between 0 and 1.[1] It is natural to assume that a bid is a strictly increasing function in valuation, and the function is invertible—if v is the valuation, then $f(v)$ is the bid function; and knowing the bid, we can compute the value using the inverse function $v = g(b) = f^{-1}(b)$.

The bidders are expected utility maximizers, in the spirit of von Neumann and Morgenstern theory. Bidder 1 with valuation v_1 will bid b_1, and will win if $b_1 > f(v_2)$. The expected payoff to bidder 1 is

$$\Pr[b_1 > f(v_2)] \cdot (v_1 - b_1) + \Pr[f(v_2) > b_1)] \cdot 0 \tag{3.1}$$

Using the inverse function $g(\cdot)$, the expression can be rewritten as

$$\Pr[g(b_1) > v_2] \cdot (v_1 - b_1) \tag{3.2}$$

---

[1] Distribution of value is *common knowledge*. Recall that information about the environment is one of the primitives discussed in Chap. 2.

As the valuations are uniformly distributed between 0 and 1, it means $\Pr[g(b_1) > v_2] = g(b_1)$. The expected payoff to bidder 1 is $g(b_1) \cdot (v_1 - b_1)$, which is maximized at $b_1^* = \frac{v_1}{2}$. Similarly, $b_2^* = \frac{v_2}{2}$. The optimal bid function is such that each bidder will bid half of her valuation and is unique. This outcome is self-enforcing, and the FPSB auction assigns the item to the bidder with the highest valuation. ◁

3.2 Mixed Strategies in Normal-form Games

In Example 3.2, we mentioned that the action sets can be extended to mixed strategies to search for outcomes that are reasonable and self-enforcing. Recall in Chap. 2, we consider players are strategic—each player strategizes to lead the game to her preferred outcome. That includes *independently* randomizing over her action set. A mixed strategy s_i is a randomization over A_i, such that Player i chooses $a_i \in A_i$ with probability $s_i(a_i)$. The set of mixed strategies is represented as S_i, and $s = (s_1, \ldots, s_n)$ is a mixed strategy profile from the set $S = S_1 \times \cdots \times S_n$. As each player randomizes independently, each action profile $a \in A$ occurs with probability $\prod_{j \in N} s_j(a_j)$. In Definition 3.1, we extend the definition of normal-form games to incorporate mixed strategies.

Definition 3.1 (*Mixed Extension of Normal-form Game*) The mixed extension of a normal-form game $\langle N, (A_i), (u_i) \rangle$ is $\langle N, (S_i), (U_i) \rangle$, where S_i is the set of probability distributions over a *finite* action set A_i, and $U_i : S \to \mathbb{R}$ is the expected utility of Player i over the outcomes induced by mixed strategy profile $s \in S$, and can be expressed as

$$U_i(s) = \sum_{a \in A} \left[\prod_{j \in N} s_j(a_j) \right] u_i(a)$$ ◀

S_i includes the strategy a_i, known as *pure strategy*, when Player i chooses a_i with probability one. It is important to note that, in Definition 3.1, we consider mixed extension only for a finite action set A_i. Considering mixed strategies for an infinite action set leads to measure-theoretic issues, which are beyond the scope of this book. We suggest Osborne and Rubinstein (1994) for different interpretations of mixed strategies.

In the following example, let us consider the mixed extension of the matching pennies game.

Example 3.5 (*Mixed Extension of Matching Pennies*) In the matching pennies game (Example 3.2), $N = \{1, 2\}$ and $A_i = \{H, T\}$. In the mixed extension, Player i randomizes the choices in A_i. The mixed strategy set of Player 1 is $S_1 = \{(p, 1 - p)\}$ $\forall p = [0, 1]$, where she chooses H with probability p, and T with probability $1 - p$. Likewise, Player 2 has a mixed strategy set $S_2 = \{(q, 1 - q)\} \ \forall q = [0, 1]$. If $p = 0.5$ and $q = 0.5$, as shown in Fig. 3.4, then each outcome occurs with probability 0.25, and the expected utilities for these mixed strategies are $U_1 = U_2 = 0.25\,[-1 + 1 - 1 + 1] = 0$.

| | | Player 2 | |
		H (0.5)	T (0.5)
Player 1	H (0.5)	−1, 1	1, −1
	T (0.5)	1, −1	−1, 1

Fig. 3.4 Mixed extension of matching pennies ◁

3.3 Two-Person Zero-Sum Games

The matching pennies game discussed in Example 3.6 is a two-person zero-sum game. Two-person zero-sum games were the initial mathematical games studied by Émile Borel, and later by John von Neumann. The *minimax theorem* for two-person zero-sum games is one of the early results, which laid the foundations of game theory and other solution concepts, is discussed in von Neumann (1928). Although this class of games has limited relevance to operations management, they are important for understanding mixed strategies, different solution concepts, and algorithms for computing the solutions (or equilibria) based on the solution concepts. The equilibrium computation problem for two-person zero-sum games can be formulated as a linear programming problem. Let us start the discussion with a variant of *Morra game*.[2]

Example 3.6 (*Matching Numbers Game*) In the matching numbers game, $N = \{1, 2\}$ and $A_i = \{3, 6\}$. Each player simultaneously announces either 3 or 6. If the numbers match, Player 1 pays Player 2 one-third of the sum of the announced numbers; otherwise, Player 2 pays Player 1 one-third of the sum of the announced numbers. This is an example of a two-person zero-sum game. Figure 3.5 shows the payoff matrix. Two-person zero-sum games are also called *matrix games* (recall bimatrix games). Player 1's payoffs are shown in Fig. 3.5. In this game, none of the four outcomes can be self-enforcing.

Assume that Player 1 independently randomizes between 3 and 6 with probability $\frac{3}{5}$ and $\frac{2}{5}$, respectively. In this case, if Player 2 announces 3, the expected payoff to Player 1 is 0, and if Player 2 announces 6, the expected payoff to Player 1 is 0.2. By playing this mixed strategy, Player 1 guarantees a non-negative payoff.

Can Player 1 improve its minimum payoff? Assume Player 1 chooses an *equalizing mixed strategy*—the expected payoff to Player 1 is the same independent of the announcement of Player 2. In the equalizing strategy, Player 1 chooses 3 and 6 with probability $\frac{7}{12}$ and $\frac{5}{12}$, respectively, with expected payoff of $\frac{1}{12}$.

---

[2] https://en.wikipedia.org/wiki/Morra_(game).

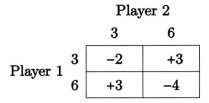

Fig. 3.5 Matching numbers game

	Player 2	
	3 ($\frac{7}{12}$)	6 ($\frac{5}{12}$)
Player 1 3 ($\frac{7}{12}$)	−2	+3
6 ($\frac{5}{12}$)	+3	−4

Fig. 3.6 Matching numbers game—minmax and maxmin strategy

Can Player 1 improve the minimum payoff further? The answer is "No" if Player 2 randomizes to minimize the maximum payment to Player 1. If Player 2 also chooses the equalizing mixed strategy that guarantees the maximum loss of $\frac{1}{12}$. Player 1's equalizing strategy is called *maxmin*, and Player 2's equalizing strategy is called *minmax* (Fig. 3.6). The equalizing strategy is also the *optimal mixed strategy*. $\frac{1}{12}$ is called the *value* of the game, and the game is biased towards Player 1.

The optimal mixed strategies are graphically illustrated in Fig. 3.7 where p is the probability of choosing 3 by Player 1 and q is the probability of choosing 3 by Player 2. In Fig. 3.7a, if Player 2 assumes that $0 \leq p < \frac{7}{12}$, then Player 2 responds by choosing 6; and if Player 2 assumes that $\frac{7}{12} < p \leq 1$, then Player 2 responds by choosing 3. Computing the responses of Player 2, Player 1 can guarantee the minimum payoff of $\frac{1}{12}$ by choosing $p = \frac{7}{12}$ that makes Player 2 indifferent between choosing 3 and 6. ($\frac{7}{12}, \frac{5}{12}$) is the optimal mixed strategy for Player 1. Similarly, in Fig. 3.7b, ($\frac{7}{12}, \frac{5}{12}$) is the optimal mixed strategy for Player 2 that bounds the maximum loss of $\frac{1}{12}$ by making Player 1 indifferent between choosing 3 and 6. The concept of making the other player indifferent to compute the optimal strategy is represented as *complementary slackness condition* in Sect. 3.3.1.

These strategies provide the basis of the most important solution concept in game theory—*Nash equilibrium*. The optimal strategies in two-person zero-sum also form Nash equilibrium (Sect. 3.4.3).

The value of the game remains the same even if we consider the payoff matrix for Player 2—Player 2 chooses the maxmin strategy, and Player 1 chooses the minmax strategy.

The mixed strategy in the matching numbers game can be implemented by using a suitable randomization device like a personal watch (Ferguson 2020). Player 1 chooses "3" if the second's hand in her watch is between 0 and 35; otherwise, choose "6". Player 2 can similarly implement his mixed strategy.

3.3 Two-Person Zero-Sum Games

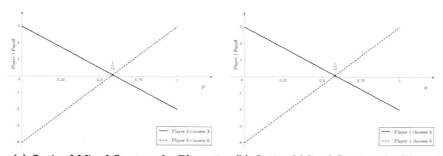

(a) Optimal Mixed Strategy for Player 1   (b) Optimal Mixed Strategy for Player 2

Fig. 3.7 Optimal mixed strategies

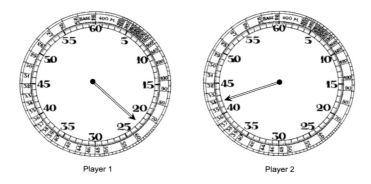

Fig. 3.8 Implementing mixed strategies

Based on the position of the second's hands in Fig. 3.8, Player 1 chooses "3" and Player 2 chooses "6". ◁

Example 3.7 Let us consider two-person zero-sum games in which the optimal strategies are pure strategies. It can be verified that the optimal strategies (maxmin and minmax) are Player 1 choosing B and Player 2 choosing L. One way of solving is to look for payoffs that are *minimum in a row* and *maximum in a column*. The value of the game is zero, and such games are called *fair games*.

		Player 2	
		L	R
Player 1	T	−2	+3
	B	0	+4

(B, L) is also the Nash equilibrium (check using the rectangle-oval method). (B, L) *is an equilibrium as unilateral deviations from it are suboptimal*—Player 1 has a lower payoff if she unilaterally decides to deviate from B to T while Player 2 continues with L. **The idea of "unilateral deviations are suboptimal" is the basis of all the solution concepts in noncooperative game theory.** ◁

3.3.1 Computing the Optimal Strategies for Two-Person Zero-Sum Games

A remarkable equivalence exists between two-person zero-sum games and linear programming[3] (von Stengel, 2022). We demonstrate the equivalence and discuss how it can be used to compute the optimal strategies. Luenberger and Ye (2008) provide an excellent introduction to linear programming.

Let us represent the $m \times n$ payoff matrix as $\mathbf{U}$, where the actions of Player 1 and Player 2 are indexed as $(1, \ldots, i, \ldots, m)$ and $(1, \ldots, j, \ldots, n)$, respectively. If Player 1 chooses action i and Player 2 chooses action j, then the payoff to Player 1 is u_{ij}. A mixed strategy for Player 1 is represented as $\mathbf{p} = (p_1, \ldots, p_i, \ldots, p_m)$, where p_i is the probability of choosing action i. Similarly, a mixed strategy for Player 2 is $\mathbf{q} = (q_1, \ldots, q_j, \ldots, q_n)$.

In any two-person zero-sum game, Player 1's objective is to maximize v (minimum guaranteed payoff) by choosing an optimal mixed strategy $\mathbf{p}^*$, independent of the choice of Player 2. Similarly, Player 2's objective is to minimize w (maximum payment to Player 1) by choosing an optimal mixed strategy $\mathbf{q}^*$, independent of the choice of Player 1. These objectives can be formulated within two linear programs that are *dual* to each other.

$$
\begin{array}{ll}
\max_{\mathbf{p}} & v \\
\text{s.t.} & \sum_{i=1}^{m} p_i u_{ij} \geq v \quad \forall j \\
& \sum_{i=1}^{m} p_i = 1 \\
& p_i \geq 0 \quad \forall i
\end{array}
\qquad
\begin{array}{ll}
\min_{\mathbf{q}} & w \\
\text{s.t.} & \sum_{j=1}^{n} u_{ij} q_j \leq w \quad \forall i \\
& \sum_{j=1}^{n} q_j = 1 \\
& q_j \geq 0 \quad \forall j
\end{array}
$$

From the *strong duality theorem* of linear programming, the optimal values of the objective functions of the dual problems are equal ($v^* = w^*$). v^* is the value of the game that is achieved when Player 1 and Player 2 use their respective optimal mixed strategies[4] $\mathbf{p}^*$ and $\mathbf{q}^*$. No player can do better by unilaterally deviating from the optimal strategy (as discussed earlier, these strategies also form Nash equilibrium).

The linear programming representation has two-fold advantages for two-person zero-sum games. Firstly, it provides the efficient *simplex algorithm* to compute the optimal strategies. Secondly, some fundamental results from the theory of linear programming enhance our understanding of the equilibrium for noncooperative games.

---

[3] This equivalence was conjectured by John von Neumann when George Dantzig was visiting Princeton University in 1947 and presented his ideas on linear programming to von Neumann.

[4] These are also called maxmin and minmax strategies for Player 1 and Player 2, respectively.

3.4 Solution Concepts for Games in Normal Form   29

At the optimal $\mathbf{p}^*$ and $\mathbf{q}^*$, and using the complementary slackness property of linear programming, the following conditions hold

$$q_j^* > 0 \Longrightarrow \sum_{i=1}^{m} p_i^* u_{ij} = v^* \qquad \clubsuit$$

$$p_i^* > 0 \Longrightarrow \sum_{j=1}^{n} u_{ij} q_j^* = v^* \qquad \spadesuit$$

Condition $\clubsuit$ means that the optimal expected payoff to Player 1 is v^* for each action j of Player 2 that is chosen with positive probability in $\mathbf{q}^*$. Condition $\spadesuit$ implies that Player 2's expected optimal payment is also v^*. Using the set of equations from Condition $\clubsuit$, Player 1 computes $\mathbf{p}^*$ to make Player 2 *indifferent* among the actions that have positive probability in $\mathbf{q}^*$. Similarly, Player 2 computes $\mathbf{q}^*$ to make Player 1 indifferent. This idea of indifference also extends to non-zero-sum noncooperative games.

Example 3.8 Let us formulate the linear programs for computing the optimal strategies for the matching numbers game.

$$
\begin{array}{ll}
\max_{\mathbf{p}} \quad v & \qquad \min_{\mathbf{q}} \quad w \\
\text{s.t.} \;\; p_1 u_{11} + p_2 u_{21} \geq v & \qquad \text{s.t.} \;\; u_{11} q_1 + u_{12} q_2 \leq w \\
\phantom{\text{s.t.}} \;\; p_1 u_{11} + p_2 u_{22} \geq v & \qquad \phantom{\text{s.t.}} \;\; u_{21} q_1 + u_{22} q_2 \leq w \\
\phantom{\text{s.t.}} \;\; p_1 + p_2 \;\; = 1 & \qquad \phantom{\text{s.t.}} \;\; q_1 + q_2 \;\; = 1 \\
\phantom{\text{s.t.}} \;\; p_1, p_2 \;\; \geq 0 & \qquad \phantom{\text{s.t.}} \;\; q_1, q_2 \;\; \geq 0
\end{array}
$$

Considering the payoff values of the game, the linear program can be written as:

$$
\begin{array}{ll}
\max_{\mathbf{p}} \quad v & \qquad \min_{\mathbf{q}} \quad w \\
\text{s.t.} \;\; -2p_1 + 3p_2 \geq v & \qquad \text{s.t.} \;\; -2q_1 + 3q_2 \leq w \\
\phantom{\text{s.t.}} \;\; 3p_1 - 4p_2 \geq v & \qquad \phantom{\text{s.t.}} \;\; 3q_1 - 4q_2 \leq w \\
\phantom{\text{s.t.}} \;\; p_1 + p_2 \;\; = 1 & \qquad \phantom{\text{s.t.}} \;\; q_1 + q_2 \;\; = 1 \\
\phantom{\text{s.t.}} \;\; p_1, p_2 \;\; \geq 0 & \qquad \phantom{\text{s.t.}} \;\; q_1, q_2 \;\; \geq 0
\end{array}
$$

Solving the above LPP, the value of the game is $v^* = w^* = \frac{1}{12}$. The optimal mixed strategy for Player 1 has $p_1^* = \frac{7}{12}$ and $p_2^* = \frac{5}{12}$. The optimal mixed strategy for Player 2 has $q_1^* = \frac{7}{12}$ and $q_2^* = \frac{5}{12}$.   ◁

3.4 Solution Concepts for Games in Normal Form

In the previous section, two-person zero-sum games provide a mathematical framework for modeling strict competition between rational, self-interested, and strategic players, with crisp results on optimal strategies (maxmin and minmax strategies). However, real strategic settings are more complex with many players and without strict competition. In this section, we discuss solution concepts for games in normal form beyond two-person zero-sum games. Prisoners' Dilemma (Example 3.1) is an example of a non-zero-sum game.

3.4.1 Pareto Optimality

Which outcomes in a game can be reasonable from an outsider's perspective? The answer to this question depends on the value judgment of the outsider. For example, if the outsider's value judgment is based on Mahatma Gandhi's Talisman[5] (Fig. 3.9), then in the following bimatrix game, if Player 2 is a weaker player, then outcome (0, 11) is reasonable from the outsider's perspective.

		Player 2	
		L	R
Player 1	T	10, 10	0, 11
	B	11, 0	3, 3

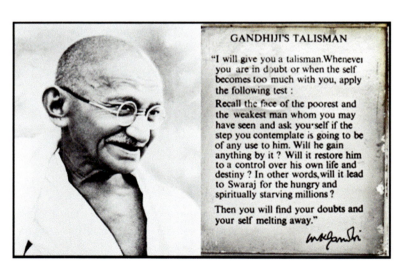

Fig. 3.9 Mahatma Gandhi's talisman

---

[5] The idea behind Mahatma Gandhi's Talisman is akin to the term *maximin* in John Rawl's "Theory of Justice" published in 1971 (Rawls 1971).

3.4 Solution Concepts for Games in Normal Form 31

Furthermore, as discussed in Sect. 2.1, utility is derived from preferences, and there is no unique utility function that explains the preferences.[6] From an outsider's perspective, the concept of *Pareto optimality*[7] provides a reasonable solution concept.

In noncooperative games, each strategy profile gives a unique outcome. To explain Pareto optimality, we consider only the pure strategy set A_i, and the associated set of pure strategy profiles $A = A_1 \times \cdots \times A_n$. However, the ideas extend easily to mixed strategy profiles $s \in S$.

Definition 3.2 (*Pareto Domination*) Strategy profile $a \in A$ *Pareto dominates* strategy profile $a' \in A$ if $u_i(a) \geq u_i(a')$ for each i, and with strict inequality for at least one i. ◀

Definition 3.3 (*Strict Pareto Domination*) Strategy profile $a \in A$ *strictly Pareto dominates* strategy profile $a' \in A$ if $u_i(a) > u_i(a')$ for each i. ◀

Definition 3.4 (*Pareto Optimality*) Strategy profile $a \in A$ is *Pareto optimal* (or *Pareto efficient*) if no other $a' \in A$ Pareto dominates a. ◀

In the above game, outcomes $(10, 10)$, $(11, 0)$, and $(0, 11)$ associated with different pure strategy profiles are Pareto optimal. $(10, 10)$ Pareto dominates $(3, 3)$, and hence $(3, 3)$ is not Pareto optimal. Recall from Example 2.1, if the actions of players decide the outcomes, the reasonable outcome is $(3, 3)$. This outcome is the only self-enforcing outcome. This outcome is also the Nash equilibrium (check the property of unilateral deviations from Nash equilibrium are suboptimal). This brings us to the classical paradox—*socially desirable outcomes may not be self-enforcing*.

Figure 3.10 illustrates the concept of Pareto domination and optimality. Outcome a Pareto dominates a'. Pareto dominance provides *partial ordering* as Pareto optimal outcomes are not comparable.

3.4.2 Domination

We move beyond the reasonable outcomes from an outsider's perspective and consider outcomes under solution concepts that are reasonable and self-enforcing from the players' perspective. One such solution concept is *dominant strategy equilibrium* based on the domination of strategies. To explain the concept of domination, we consider only pure strategy sets A_i; however, the concept extends to mixed strategy sets S_i.

Before we go further, a standard convention in game theory is to represent the strategy profile $(a_1, \ldots, a_i, \ldots a_n)$ **as** (a_i, a_{-i}) **if the focus is on the choice of player** i. a_{-i} **is the strategy profile of all the other players (excluding** i**).** $a_{-i} \in A_{-i} = A_1 \times \cdots \times A_{i-1} \times A_{i+1} \times \cdots \times A_n$.

---

[6] To circumvent this problem, in coalitional games with transferable utility (Sect. 2.2.3), it is assumed that the utilities are measured in the same currency and are transferable.

[7] Pareto optimality is named after Italian civil engineer and economist Vilfredo Pareto (1848–1923), who used the concept for defining economic efficiency.

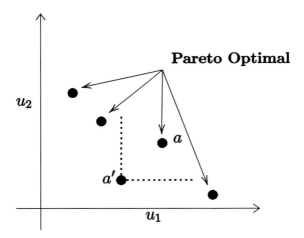

Fig. 3.10 Pareto domination and optimality

Definition 3.5 (*Strict Domination*) Strategy a_i *strictly dominates* strategy a'_i for Player i if $u_i(a_i, a_{-i}) > u_i(a'_i, a_{-i})$ for each $a_{-i} \in A_{-i}$. ◂

Definition 3.6 (*Weak Domination*) Strategy a_i *weakly dominates* strategy a'_i for Player i if $u_i(a_i, a_{-i}) \geq u_i(a'_i, a_{-i})$ for each $a_{-i} \in A_{-i}$, and with strict inequality for at least one $a_{-i} \in A_{-i}$. ◂

Definition 3.7 (*Dominant Strategy Equilibrium*) A strategy $(a_1, \ldots a_n)$ forms a *dominant strategy equilibrium* if every a_i in the profile is a dominant strategy (strict or weak) for Player i. ◂

Every dominant strategy equilibrium is also a Nash equilibrium; however, the converse is not true. A strictly dominant strategy equilibrium is necessarily the unique Nash equilibrium. Dominant strategy equilibrium is the most desired solution concept when the games are designed. We will discuss the game design problem as *mechanism design* in Chap. 9.

Let us revisit the Prisoners' Dilemma and solve it by domination.

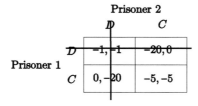

Example 3.9 (*Domination in Prisoners' Dilemma*) C is the *strictly dominant strategy* for each prisoner, irrespective of the choice made by the other prisoner. (C, C) is the strictly dominant strategy equilibrium and also the unique Nash equilibrium. ◂

3.4 Solution Concepts for Games in Normal Form   33

The above example illustrates the idea of computing Nash equilibrium by eliminating dominated strategies (D in Prisoners' Dilemma). A strategy is *dominated* if any other strategy (strictly or weakly) dominates it. It is reasonable to assume that each player can deduce that other players do not choose a dominated strategy and can be eliminated when deciding the best strategy. The order of eliminating strictly dominated strategies does not affect the final outcomes (known as *Church-Rosser property*); however, it does not hold while eliminating weakly dominated strategies (the order of eliminating weakly dominated strategies affects the final outcome).

Example 3.10 (*Second-Price Sealed-Bid Auction*) We discussed first-price sealed-bid (FPSB) auction in Example 3.4 where bidders bid simultaneously in sealed bids for an item, and the bidder with the highest bid wins the auction and pays her own bid to get the item. In second-price sealed-bid (SPSB) auction,[8] bidders bid simultaneously in sealed bids for an item, and the bidder with the highest bid wins the auction and pays the second-highest bid to get the item. In SPSB auction, *bidding truthfully* (or $b_i = v_i$) *is a dominant strategy for each bidder* $i \in N$.

Let us consider SPSB auction with two bidders who are privately informed of their valuation for an item. Bidder 1 with valuation v_1 bids b_1, and Bidder 2 with valuation v_2 bids b_2. If $b_1 \geq b_2$, Bidder 1 wins the auction, and gets a payoff of $v_1 - b_2$. If $b_1 < b_2$, Bidder 1 loses the auction, and gets a payoff of 0. If Bidder 1 bids more than her valuation ($b_1 > v_1$), then there is a possibility of a negative payoff when $b_1 > b_2 > v_1$, which can be avoided if Bidder 1 bids truthfully and loses. If Bidder 1 bids less than her valuation ($b_1 < v_1$), then there is a possibility of payoff of 0 when $b_1 < b_2 < v_1$. In this case, Bidder 1 gets a payoff of $v_1 - b_2$ by bidding truthfully and winning the auction. Hence, bidding her valuation is a dominant strategy for Bidder 1. This analysis is valid for Bidder 2, and can be extended to SPSB auction with n bidders. We discuss FPSB and SPSB auctions as part of mechanism design in Chap. 9.   ◁

3.4.3 Nash Equilibrium

Nash equilibrium is the most important solution concept for noncooperative games. Kreps (1990) mentions that the Nash outcomes are the most likely outcomes to be observed in a game when the players are rational, self-interested, and strategic. We have done some hand-waving about Nash equilibrium in the previous sections. In Example 3.1, the best responses were identified using the rectangle-oval method, and the intersection of the best responses is Nash equilibrium. In this section, we concretize the idea of Nash equilibrium for games in normal form; however, Nash equilibrium has been refined into other solution concepts for games in extensive form. *Subgame perfect Nash equilibrium* is one such solution concept (refer Sect. 5.3.1).

---

[8] SPSB auction is also known as *Vickery auction*, named after William Vickery, who was awarded the 1996 Nobel Memorial Prize in Economic Sciences for using game theory in auctions. Vickrey (1961) is one of his seminal papers on auction theory.

We start with some definitions. We use mixed strategies to define Nash equilibrium, as some interesting results on Nash equilibrium need mixed strategies.

Definition 3.8 (*Best Response*) Mixed strategy $s_i^* \in S_i$ is the best response of Player i to the mixed strategy profile s_{-i} if $u_i(s_i^*, s_{-i}) \geq u_i(s_i, s_{-i})$ for each $s_i \in S_i$. ◀

The best response may not be unique. If s_i^* includes two or more actions, then any other mixing of the same actions will also be the best response. This can be illustrated in the matching numbers game. In Fig. 3.6, Player 1 has the optimal mixed strategy—choosing 3 with probability $\frac{7}{12}$ and 6 with probability $\frac{5}{12}$, with the expected payoff $\frac{1}{12}$. It can be easily verified that if Player 2 continues to randomize with probabilities $\frac{7}{12}$ and $\frac{5}{12}$, any other probability mixing of 3 and 6 by Player 1 has the same expected payoff of $\frac{1}{12}$.

As mentioned earlier, the intersection of the best responses is Nash equilibrium, which is formally stated in the following definition.

Definition 3.9 (*Nash Equilibrium*) Strategy profile $(s_1^*, \ldots, s_n^*)$ is a *Nash equilibrium* if for each agent $i \in N$, $u_i(s_i^*, s_{-i}^*) \geq u_i(s_i, s_{-i}^*)$ for each $s_i \in S_i$. ◀

It means that a strategy profile is Nash equilibrium if each player's strategy in the profile is the best response to the strategies of the other players in the profile. In other words, if other players are playing their part of Nash equilibrium, then unilateral deviation leads to a lower payoff for each Player $\Rightarrow$ *Nash equilibrium is self-enforcing*. It can be checked in the following bimatrix game where (B, R) is Nash equilibrium.

Player 2

		L	R
	T	10, 10	0, 11
Player 1	B	11, 0	3, 3

Shoham and Leyton-Brown (2009) classify Nash equilibrium into *strict Nash* and *weak Nash*. In strict Nash equilibrium, the inequality in Definition 3.9 is strict (the best response of each player is unique). Nash equilibrium is weak if it is not strict. It is intuitive that weak Nash equilibrium is *less stable* than strict Nash equilibrium. As discussed earlier, if a mixed strategy includes two or more actions, then any other mixing of the same actions will also be the best response. Hence, any mixed-strategy Nash equilibrium is always weak. This idea is used to compute mixed-strategy Nash equilibrium—each player randomizes with probabilities to make the other player indifferent (recall Conditions ♣ and ♠ used to compute the optimal strategies on Sect. 3.3.1). The following example illustrates computing mixed-strategy Nash equilibrium.

Example 3.11 (*Restaurant game*) In the restaurant game, two friends—Alpha and Beta—desire to go together for lunch at either *Indian* or *Chinese* restaurant. Alpha

3.4 Solution Concepts for Games in Normal Form 35

prefers Indian over Chinese, while Beta prefers Chinese over Indian. They decide to stay at home if they cannot coordinate the choice. The payoff matrix is shown in Fig. 3.11.

		Beta	
		Indian	Chinese
Alpha	Indian	2, 1	0, 0
	Chinese	0, 0	1, 2

Fig. 3.11 Restaurant game

The game has two pure-strategy Nash equilibria—(Indian, Indian) and (Chinese, Chinese) (computed using the rectangle-oval method). There is mixed-strategy Nash equilibrium in which Alpha chooses Indian and Chinese restaurants with probabilities p and $1 - p$, respectively, while Beta chooses Indian and Chinese restaurants with probabilities q and $1 - q$, respectively. Mixed-strategy Nash equilibrium is computed using the idea that each player randomizes to make the other player indifferent—Alpha computes p to make Beta indifferent between Indian and Chinese restaurants.

$$1 \times p + 0 \times (1 - p) = 0 \times p + 2 \times (1 - p) \Rightarrow p = \frac{2}{3}$$

Similarly, $q = \frac{1}{3}$. In mixed-strategy equilibrium, the expected payoff to each agent is $\frac{2}{3}$. If Beta sticks to mixed strategy $(\frac{1}{3}, \frac{2}{3})$, then any other mixing of pure strategies by Alpha gives the same expected payoff of $\frac{2}{3}$, which means mixed-strategy Nash equilibrium is weak. There are, in fact, infinite mixed strategies that are the best response to the mixed strategy used by other players; however, it is conjectured in game theory that each player chooses *just the right probabilities* that ensure mixed-strategy Nash equilibrium. Furthermore, each pure-strategy Nash equilibrium Pareto-dominates mixed-strategy Nash equilibrium.

An important question—which one of Nash equilibria should be selected in the restaurant game? This is one of the most challenging problems in game theory—*the equilibrium selection problem*. We suggest Harsanyi and Selten (1988) for an elaborate discussion on this problem. ◁

Example 3.12 (*Braess's paradox*) In this example, we discuss Nash equilibrium in transportation networks. Braess (1968) discusses a paradox called *Braess's paradox*, in which adding a superfast link to reduce congestion increases travel time for all travelers. Consider the network shown in Fig. 3.12. Six motorists wish to travel from Node 1 to Node 4. There are two routes $1 \to 2 \to 4$ and $1 \to 3 \to 4$. The motorists are rational and self-interested, and can decide on the route to minimize travel time. The travel time for each link in minutes is shown in Fig. 3.12, where x_{ij} is the number of motorists using link ij. In this network, there is a unique Nash equilibrium with three motorists on each route and travel time of 51 min.

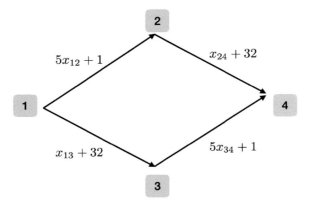

Fig. 3.12 Braess's paradox: initial network

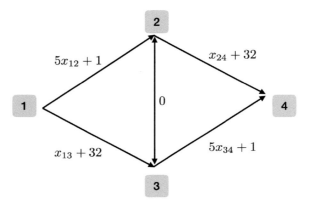

Fig. 3.13 Braess's paradox: augmented network

The government is not happy with travel time of 51 min, and provide link 23 with instant commuting[9] in both directions, as shown in Fig. 3.13. With new link 23, the motorists prefer to choose $1 \to 2 \to 3 \to 4$ with travel time of 62 min, which is higher than 51 min in the initial network. It can be verified that this is a unique Nash equilibrium, as unilateral deviations are suboptimal for any motorist.

We discuss Braess's paradox in detail in Sect. 4.2. ◁

3.4.3.1 Existence of Nash Equilibrium

As discussed earlier, using mixed strategies leads to weak Nash equilibrium; however, they are needed for the existence of Nash equilibrium.

---

[9] Instant commuting is possible using *hyperloop* transportation system (refer https://en.wikipedia.org/wiki/Hyperloop).

3.4 Solution Concepts for Games in Normal Form

Theorem 3.1 (Existence of Nash Equilibrium; Nash, 1951) *For every finite game, there exists at least one Nash equilibrium in mixed strategies.* ◀

Proof We discuss the idea behind the proof of Theorem 3.1 using Kakutani fixed-point theorem, which is stated as *Let S be a nonempty, convex, bounded, and closed subset of a finite-dimensional vector space, and $F : S \rightrightarrows S$ is a correspondence. If the graph of F is closed for every $s \in S$ and $F(s)$ is a convex and nonempty set, then there exists an $s^* \in S$ such that $s^* \in F(s^*)$* (Kakutani 1941).[10]

S_i is the set of mixed strategies; hence convex, closed, bounded, and nonempty. $S = S_1 \times S_2 \times ... \times S_n$, it is obvious that S is a non-empty, convex, bounded, and closed subset of Euclidean space. Hence, it is left to prove that $F(s)$ is a non-empty, convex set, and its graph is closed. Breaking down the correspondence into smaller fragments, define $F_i(s) = \{\hat{s}_i \in S_i : \hat{s}_i$ is best response to $s_{-i}\}$. Since pure strategies are a special case of mixed strategies, $F_i(s)$ is non-empty as there will be at least one strategy that is the best response to other players' strategies. Convexity is directly implied for a unique best response. With multiple best responses, convexity still holds as any other mixing of pure strategies yields the same payoff (refer Example 3.11).

To prove $F_i(s)$ is a closed graph, it suffices to prove that $F_i(s)$ is upper semi-continuous. Since $F_i(s)$ is a collection of best responses, any type of convex combination of best responses is also the best response. Therefore, if $F_i(s)$ has two images, these two images are connected through mixed strategies. Moreover, every subsequence converges within the space of $F_i(s)$. Given this upper semi-continuous property, which essentially implies closed graph property of $F_i(s)$, $F(s)$ forms a closed graph as $F(s) = F_1(s) \times F_2(s) \times ... \times F_n(s)$. As a result of implications, there exists a fixed point and $s^* \in F(s^*)$. In simple terms, a finite game in mixed strategies guarantees all the prerequisites of containing a fixed point, which provides proof of Nash equilibrium's existence. □

3.4.3.2 Computation of Nash Equilibria

In Sect. 3.3.1, we have seen that the problem of computing the optimal strategies that form Nash equilibria in a two-person zero-sum game can be formulated as a linear programming problem, and the efficient simplex algorithm can be used to compute the optimal strategies. The problem of computing Nash equilibria for non-zero-sum games has a higher complexity. Computing Nash equilibria in two-person

---

[10] The theorem needs basic familiarity with the concepts from real analysis. Pugh (2002) provides an excellent introduction to real analysis. Let X denote a subset of the finite-dimensional vector space $\mathbb{R}^n$.

Convex set: X is convex if and only if, for every two vectors $x, x' \in X$ and for each $\lambda \in [0, 1]$, $\lambda x + (1 - \lambda)x' \in X$.

Bounded set: X is bounded if and only if there exists a positive number M such that for each $x \in X$, $\sum_{i=1}^{n} | x_i | \leq M$

Closed set: A set X is closed if and only if every convergent sequence converges in X.

Correspondence: Y is also a subset of the finite-dimensional vector space $\mathbb{R}^n$. A correspondence or a set function $G : X \rightrightarrows Y$ assigns for each $x \in X$, a subset of Y, that is $G(x) \subseteq Y$.

Player 2

0,0	6,2	0,-4
2,4	0,0	2,6
-6,4	8,2	10,0

Fig. 3.14 3×3 bimatrix game

non-zero-sum games can be formulated as a *linear complementarity problem (LCP)*, and *Lemke-Howson algorithm* (Lemke & Howson, 1964) is one of the well-known algorithms for solving LCPs.

In this section, we discuss the geometric representation of the best-response regions, identifying Nash equilibria, and the Lemke-Howson algorithm. The geometric representation enhances understanding of the best response structure of Nash equilibria and also provides an alternative proof of the existence of Nash equilibrium in bimatrix games. The discussion is primarily based on Shapley (1974), Shoham and Leyton-Brown (2009), and von Stengel (2021).

Let us consider the following two-person non-zero-sum game (3×3 bimatrix game) in Fig. 3.14. The goal is to identify Nash equilibria of the bimatrix game and discuss the Lemke-Howson algorithm to find one Nash equilibrium of the bimatrix game.

Best-response Regions

Step I: **Labeling Pure Strategies**: In this step, each pure strategy is given a unique label. For example, in $m \times n$ bimatrix game, Player 1 pure strategies are labeled $I = \{1, \ldots, m\}$ and Player 2 pure strategies are labeled $J = \{m + 1, \ldots, m + n\}$. This is shown in Fig. 3.15. Each pure strategy is assigned a unique number label as well as a unique color label. The color labeling is to improve exposition.

Step II: **Representing Mixed Strategies:** In this step, mixed strategies are represented as a convex combination of pure strategies as shown in Fig. 3.16. Mixed strategy x for Player 1 is a three-dimensional vector (x_1, x_2, x_3) that assigns probability x_i to pure strategy $i \in I$. The set of mixed strategies forms *simplex X*, which is represented as the shaded plane in 3.16. Mixed strategy y for Player 2 is a three-dimensional vector (y_4, y_5, y_6) that assigns probability y_j to pure strategy $j \in J$. The set of mixed strategies for Player 2 is represented as simplex Y. It is important to observe that the axes are labeled based on the labels used in Step I.

Step III: **Identifying and Labeling the Best-Response Regions:** The objective is to identify the best response regions in simplex Y for each pure strategy of Player 1, as the best response to the mixed strategy of Player 2.

3.4 Solution Concepts for Games in Normal Form

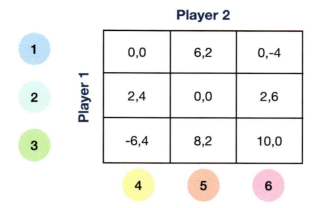

Fig. 3.15 Labeled pure strategies

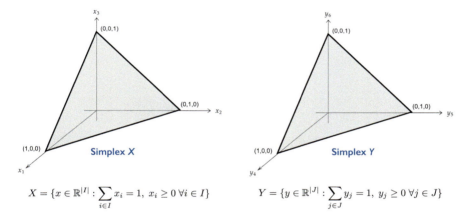

Fig. 3.16 Mixed strategy representation

For the bimatrix game in Fig. 3.14, the expected payoffs of Player 1 for each pure strategy if Player 2 chooses mixed strategy y are

① $6y_5$

② $2y_4 + 2y_6$

③ $-6y_4 + 8y_5 + 10y_6$

Pure strategy ① is the best response to y if the expected payoff of ① is at least as high as of ② and ③, which implies

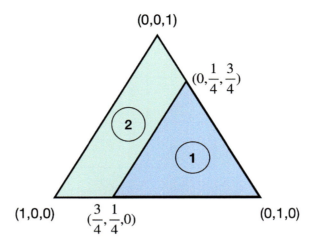

Fig. 3.17 Best response regions $Y\text{①}$ and $Y\text{②}$

$$6y_5 \geq 2y_4 + 2y_6 \text{ and } 6y_5 \geq -6y_4 + 8y_5 + 10y_6$$

on rearranging
$$3y_5 \geq y_4 + y_6 \text{ and } 3y_4 \geq y_5 + 5y_6$$

The best response region in simplex Y for pure strategy $i \in I$ of Player is $Y\text{①} = \{y \in Y : i \text{ is the best response to } y\}$. The best response regions $Y\text{①}$ and $Y\text{②}$ by comparing strategy ① and strategy ② are shown in Fig. 3.17. $Y\text{①}$ is the set of mixed strategies y with label ①. The line segment joining points $(\frac{3}{4}, \frac{1}{4}, 0)$ and $(0, \frac{1}{4}, \frac{3}{4})$ is the set of mixed strategies y for which both ① and ② are the best responses, hence has two labels.

Similarly, pairwise comparisons between ① and ③, and between ② and ③ lead to the best regions as shown in Fig. 3.18a and Fig. 3.18b, respectively.

The intersection of the pairwise best response regions is shown in Fig. 3.19. The best response regions are labeled based on the pure strategies of Player 1.

Step IV: Representing Own Pure Strategies: In this step, mixed strategy y is also labeled with Player 2's pure strategies that are not chosen with positive probability in y—use label $j \in J$ for $Y\text{①} = \{y \in Y : y_j = 0\}$. These additional labels on simplex Y are shown in Fig. 3.20. The edge joining $(1, 0, 0)$ and $(0, 0, 1)$ is labeled ⑤, as pure strategy ⑤ is chosen with zero probability ($y_5 = 0$) in any mixed strategy on edge joining $(1, 0, 0)$ and $(0, 0, 1)$. The corner point $(1, 0, 0)$ has labels ⑤ and ⑥ as $y_5 = y_6 = 0$. It is important to note that all the pure strategy labels of the bimatrix game in Fig. 3.15 are shown in simplex Y.

Fully-labeled simplices X and Y are shown in Fig. 3.21. The dots indicate points that have three labels.

3.4 Solution Concepts for Games in Normal Form 41

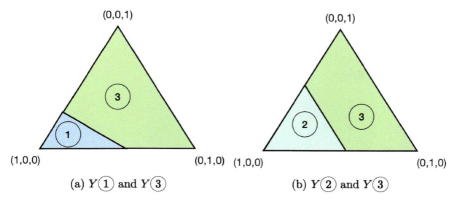

Fig. 3.18 Pairwise best response regions

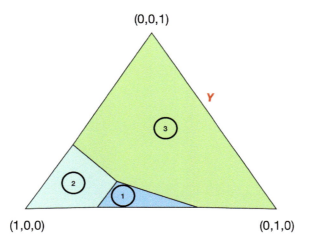

Fig. 3.19 Best response regions in simplex Y

Identifying Nash Equilibria

Nash equilibrium of a bimatrix game is a pair of mixed strategies (x, y) that are best responses to each other. From Definition 3.8, the best responses are not unique, and any pure strategy that has a positive probability in mixed strategy is also the best response. In other words, in a mixed strategy, each pure strategy is either the best response (labeled in the simplex of the other player) or has zero probability (labeled in its own simplex). Mixed strategies are in equilibrium if pure strategy with positive probability is the best response to the other player's mixed strategy and vice versa. It implies that *a pair of mixed strategies in equilibrium has all the labels*. For a bimatrix game $m \times n$, the equilibrium pair has each of $m + n$ labels. To identify Nash equilibria, we search for a pair of mixed strategies that have all the labels. This means we restrict our search to points $x^1, \ldots, x^7$ and $y^1, \ldots, y^7$, as each of these points has three labels.

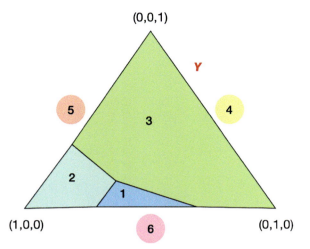

Fig. 3.20 Labeled Player 2's pure strategies in simplex Y

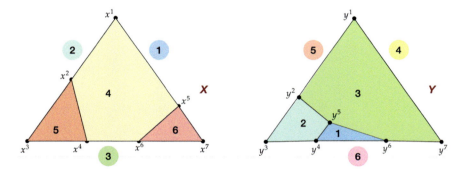

Fig. 3.21 Fully-labeled simplices X and Y

Let us start with the pair (x^1, y^3) to check whether it forms equilibrium. y^3 is the best response to x^1 as y^3 only uses pure strategy ④ with a positive probability and x^1 has label ④ in the best response region. However, x^1 is not the best response to y^3 as x^1 uses pure strategy ③ with a positive probability, and y^3 does not have label ③ in the best response region. The pair (x^1, y^3) is not fully labeled, as label ③ is missing. It can be verified that three pairs form Nash equilibria, as shown in Fig. 3.22. Each mixed strategy pair in equilibrium has all six labels (number labels as well as color labels).

Lemke-Howson algorithm

We now discuss the Lemke-Howson algorithm to find one Nash equilibrium. Lemke-Howson algorithm searches for a fully labeled pair by following a path through pairs (x, y) in simplices X and Y.

The algorithm starts by adding points x^0 in which each pure strategy $i \in I$ is assigned zero probability, and hence all labels in I; and y^0 in which each pure

3.4 Solution Concepts for Games in Normal Form

Fig. 3.22 Nash equilibria

strategy $j \in J$ is assigned zero probability, and hence all labels in J. (x^0, y^0) forms the artificial equilibrium pair as it is fully labeled.

In the next step, the algorithm selects one of the simplices, allows missing one of the labels, and moves to the point with the missing label in the selected simplex. This is illustrated in Fig. 3.23, where simplex X is selected, and we assume that ① is the missing label, then x^3 has label ① missing. The algorithm moves from x^0 to x^3 that has labels ②, ③, and ⑤. The new pair (x^3, y^0) is almost fully labeled with ① missing and ⑤ appearing twice.

As label ⑤ appeared twice, in the next step, the algorithm moves from y^0 to point y^7 in simplex Y that has label ⑤ missing. The new pair (x^3, y^7) is almost fully labeled with ① still missing and ③ appearing twice. Using similar arguments, the algorithm next moves to (x^2, y^7) that is almost fully labeled with ① still missing and ④ appearing twice. In the final step, the algorithm drops label ④ and moves to (x^2, y^6). (x^2, y^6) is fully labeled, and hence is Nash equilibrium (also refer Fig. 3.22).

In Fig. 3.23, a point can be uniquely defined using three labels. In the Lemke-Howson algorithm, dropping label ① leads to the particular point x^3 in simplex X with a unique and connected path from x^0. This step leads to the particular point y^7 in simplex Y to drop the duplicate label ⑤ through a unique and connected path from y^0. As there are only finite pairs to check, the algorithm must terminate at an equilibrium following a unique path from the artificial equilibrium. The unique path is shown with blue arrows in Fig. 3.23. The algorithm also provides an alternative proof of the existence of Nash equilibrium in bimatrix games.

It is important to note that the Lemke-Howson algorithm is *nondeterministic*, though the nondeterminism is confined in the initial move, and the algorithm follows a unique path in the subsequent steps before converging to an equilibrium. For illustration, if we assume label ② as the missing label in the first step, then the path traced to another equilibrium (x^5, y^2) from (x^0, y^0) is shown in Fig. 3.24.

One of the drawbacks of the Lemke-Howson algorithm is that it does not guarantee finding all Nash equilibrium. Furthermore, there are instances when the worst-case behavior of the algorithm has exponential time complexity. Chen et al. (2009) show that the problem of finding Nash equilibrium of a bimatrix game belongs to a complexity class called **PPAD** (**P**olynomial **P**arity **A**rgument, **D**irected version). We recommend Shoham and Leyton-Brown (2009) and Roughgarden (2016) for a detailed discussion on **PPAD**.

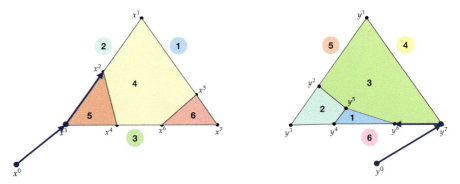

Fig. 3.23 Unique path to equilibrium (x^2, y^6) from artificial equilibrium (x^0, y^0)

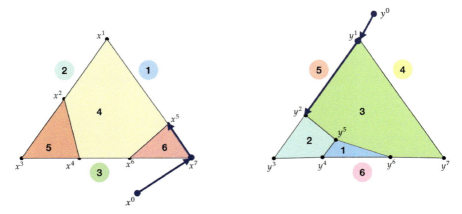

Fig. 3.24 Unique path to equilibrium (x^5, y^2) from artificial equilibrium (x^0, y^0)

3.4.4 Preplay Communication and Correlated Equilibrium

3.4.4.1 Cheap Talk

As discussed in Chap. 2, communication is modeled in noncooperative games, which also includes *preplay communication* (or *cheap talk*). In "Golden Balls" (Example 2.2), the players have to choose between *steal* or *split*. In the episode, the presenter Jasper Carrott allowed preplay communication between the players Abraham and Nick, and Nick declared "steal" during the preplay communication. The declaration is costless to Nick and hence cheap talk. Similarly, in Example 2.1, during preplay communication, if Player 1 declares "I will choose T" and Player 2 declares "I will choose L", then the communication is not just cheap but also non-credible. The agreed-upon outcome is not self-enforcing. Players can sign binding contracts to make preplay communication credible (refer Sect. 4.4). In recent years, technological solutions like *blockchains* have been used for credible communication among the agents (refer Sect. 4.5).

3.4 Solution Concepts for Games in Normal Form

Fig. 3.25 Traffic Chaos on Dagen H (*Source* https://en.wikipedia.org/wiki/Dagen_H)

Preplay communication can allow players to invent *conventions* and coordination mechanisms to play the game. Driving on a road's left or right-hand side is a convention that coordinates traffic. This convention is different in different countries—India follows driving on the left hand, while in the United States, it is driving on the right hand. One interesting episode is *Dagen H*—on September 3, 1967, Sweden changed the convention from the left to the right-hand side (Fig. 3.25). Iceland also changed from the left-hand to right-hand drive on May 26, 1968 (*H-dagurinn*). We recommend Lewis (2002) and Chwe (2013) for game-theoretic discussion on conventions.

3.4.4.2 Correlated Equilibrium

During preplay communication, players can devise coordination mechanisms that recommend strategies to each player such that unilateral deviation from the recommendations is suboptimal for each player. Aumann (1974) proposes a solution concept for noncooperative games, which includes such possibilities, is *correlated equilibrium*, a generalization of Nash equilibrium.

For example, in the restaurant game (Example 3.11), during preplay communication, Alpha and Beta can devise the following mechanism to coordinate their strategy—"toss an unbiased coin, choose Indian if the head and Chinese if tails" (Fig. 3.26). In this mechanism, the strategies are correlated, and it can be verified that unilateral deviation from the recommended strategy is suboptimal. Such correlated strategies for which unilateral deviation is suboptimal (hence self-enforcing) form correlated equilibrium. The expected payoff is $\frac{3}{2}$, and it Pareto-dominates the expected payoff $\frac{2}{3}$ in mixed-strategy equilibrium. The *unbiased coin* can be replaced by an *unbiased referee* who can toss the coin and, based on the outcome, communicate only the recommended strategy to the players. Using such mechanisms, any payoff in the convex combination of Nash equilibria payoffs can be achieved as cor-

46   3 Games in Normal Form

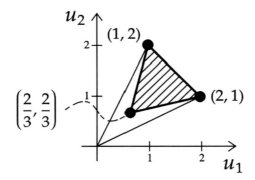

Fig. 3.26 Correlated strategies in restaurant game

Fig. 3.27 Set of correlated equilibrium in restaurant game

related equilibrium, shown as the shaded area in Fig. 3.27. Let us consider another example to discuss correlated equilibrium.

Example 3.13 (*Gig-economy game*) Consider a simple setting of a gig economy where two workers, A and B, work towards a goal by simultaneous and joint effort. If both the workers choose high effort "H", each gets the payoff of 2. However, if both choose low effort "L", then the goal is not achieved, both are penalized, and the payoff is -1 to each. If one chooses "H" and the other chooses "L", the player with the choice of "L" can free-ride[11] on the other worker and has the payoff of 3. The payoff matrix is shown in Fig. 3.28. The game has two pure Nash equilibria with payoffs (3, 0) and (0, 3), and one mixed Nash equilibrium with payoff (1, 1). As mentioned earlier, any payoff in the convex combination of Nash equilibria payoffs can be achieved as correlated equilibrium. For example, during preplay communication, workers can decide to coordinate on pure-strategy Nash equilibrium based on the outcome of an unbiased coin, with the expected payoff of $1\frac{1}{2}$.

Can the workers devise a mechanism during preplay communication that achieves a payoff outside the convex combination of Nash equilibria payoffs as correlated equilibrium? In one such mechanism, the workers agree to take the assistance of an unbiased referee. The referee will roll an unbiased six-sided dice, and based on

---

[11] This can happen in any moral hazard scenario when the effort is not observable. Technology solutions like Internet-of-Things (IoT) have been increasingly used to mitigate this problem.

3.4 Solution Concepts for Games in Normal Form

		Worker B	
		H	L
Worker A	H	2, 2	0, 3
	L	3, 0	−1, −1

Fig. 3.28 Payoffs in gig-economy game

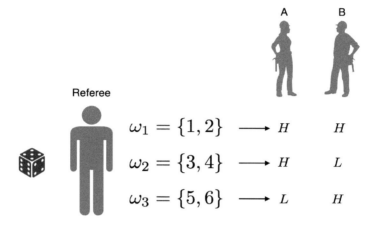

Fig. 3.29 Correlated equilibrium in gig-economy game

the outcome, *privately* communicates the recommended strategy to each worker, as shown in Fig. 3.29.

From Fig. 3.29, it is obvious that if worker A is recommended to choose H, she knows that either state ω_1 or ω_2 has occurred but cannot distinguish between them. She can assign conditional probabilities of $\frac{1}{2}$ to ω_1 and ω_2. She can also infer that worker B has been recommended to choose H if state ω_1 has occurred or L if state ω_2 has occurred. In this case, the expected payoff to worker A by following the recommendation is $\frac{1}{2} \times 2 + \frac{1}{2} \times 0 = 1$, which is the same even if worker A rejects the recommendation and chooses L (worker A gains nothing by deviation). If worker A is recommended to choose L, she knows that state ω_3 has occurred, and worker B has been recommended to choose H. In this case, worker A's payoff is 3 if she follows the recommendation that is higher than 2 if she decides to deviate to H. The analysis is similar for worker B.

Correlated strategies for each state are shown in Fig. 3.30. As each state can occur with probability $\frac{1}{3}$, the expected payoff in this correlated equilibrium is $\left(1\frac{2}{3}, 1\frac{2}{3}\right)$. It can be verified that the expected payoff vector lies outside the convex combination of Nash equilibria payoffs and Pareto-dominates mixed-strategy Nash equilibrium. ◁

As evident from Fig. 3.26, the set of correlated equilibria is infinite, and mixed-strategy Nash equilibrium also belongs to that set. One advantage is that it is easier to compute than Nash equilibrium, as the problem of computing correlated equilibrium can be formulated as a *linear feasibility problem*.

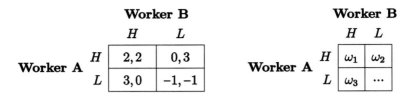

Fig. 3.30 Correlated strategies in gig-economy game

The formal definition of correlated equilibrium is

Definition 3.10 (*Correlated Equilibrium*) For game $\langle N, (A_i), (u_i) \rangle$, correlated equilibrium consists of

- A finite probability space (Ω, π)
- Information partition $\mathcal{P}_i$ of Ω for each player
- Strategy functions $s_i : \Omega \mapsto A_i$, with $s_i(\omega) = s_i(\omega')$ when $\omega, \omega' \in P_i$ for some $P_i \in \mathcal{P}_i$
- s'_i is another strategy function

then

$$\sum_{\omega \in P_i} \pi(\omega) u_i(s_i(\omega), s_{-i}(\omega)) \geq \sum_{\omega \in P_i} \pi(\omega) u_i(s'_i(\omega), s_{-i}(\omega)) \quad \blacktriangleleft$$

In the above definition, the probability space and information partition are part of the equilibrium. The set of correlated equilibria includes mixed-strategy Nash equilibrium (refer to Fig. 3.27); it can be deduced using Theorem 3.1 that for a finite game, correlated equilibrium always exists.

3.4.5 Bayesian-Nash Equilibrium

In Chap. 2, we mentioned that one of the primitives in game theory is the information about the environment in which a game is embedded. In *games of complete information*, the information about the environment is complete. For example, in Example 3.11, each player knows the action set and payoffs of the other player, and only the choice of the other player is not known to each player. Different solution concepts, like Nash equilibrium, provide reasonable outcomes for games of complete information. In Example 3.13, the unbiased referee coordinates the action of the workers using correlated equilibria in a game of complete information. However, in many real settings, players have incomplete information about the other players' payoffs,[12] and such games are called *games of incomplete information*.

In games of incomplete information, players are unsure about *which game is being played*? Let us illustrate the incomplete information setting through an example.

---

[12] Incomplete information on other attributes like the number of players or their action sets can be reduced to incomplete information about payoffs (Shoham & Leyton-Brown, 2009).

3.4 Solution Concepts for Games in Normal Form

Example 3.14 (*Vaccine Procurement Problem*)[13] Consider a procurement setting where a government has to procure 100 million vaccine doses. Firm A and Firm B are the only two manufacturers of the vaccine. The cost of producing a single dose is not publicly observable and is private information for each firm. However, it is common knowledge that the cost per dose in dollars can be either *one* or *two*, and *each cost is equally likely*. The setting is illustrated in Fig. 3.31. In this setting, the environment has four possible *states*—$(C_A = 1, C_B = 1)$, $(C_A = 1, C_B = 2)$, $(C_A = 2, C_B = 1)$, $(C_A = 2, C_B = 2)$

Government has to procure 100 million vaccine doses at lowest possible cost

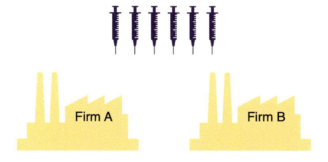

$C_i = \{1, 2\}$ in dollars. Each cost is equally likely.
$i = A, B$
Cost information is private information

Fig. 3.31 Vaccine procurement

The government uses a mechanism shown in Fig. 3.32. In the mechanism, the government asks them to reveal their costs by choosing an action $m_i \in \{1, 2\}$, which messages each firm's cost to the government. The rules include simultaneous choice of actions and the procurement scheme based on the choice of actions. As shown in Fig. 3.32, the procurement scheme is—if they message the same cost, the order is split, and each firm is paid the messaged cost; and if their messages differ, the complete order is given to the firm with lower cost, and paid its messaged cost. For example, if $m_A = 1$ and $m_B = 2$, then the procurement scheme orders 100 million doses from firm A ($x_A(1, 2) = 100$; $x_B(1, 2) = 0$) at a price $t = 1$ dollar per dose.

---

[13] This problem is discussed in detail as a mechanism design problem in Chap. 9.

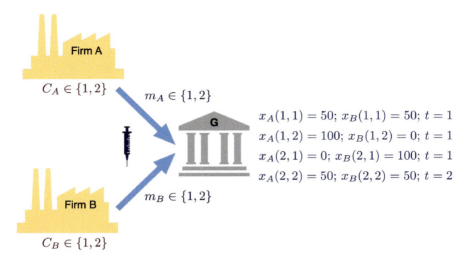

Fig. 3.32 Vaccine procurement mechanism

The game of incomplete information associated with the government's procurement mechanism is shown in Fig. 3.33. The environment has four possible states, each with a probability of 0.25, and *each state leads to a different bimatrix game* in Fig. 3.33. The payoffs are computed based on the choice of m_A and m_B, given their respective costs. For example, in the game with $C_A = 1$, $C_B = 1$, if both the firms message 2, then the order is divided, and each makes a profit of 50 million dollars. The red boxes are the information partition of Firm A, and the blue boxes are the information partition of Firm B. The top red box is the information partition of Firm A when $C_A = 1$.

	$m_B = 1$	$m_B = 2$
$m_A = 1$	0, 0	0, 0
$m_A = 2$	0, 0	50, 50

$C_A = 1, C_B = 1$

	$m_B = 1$	$m_B = 2$
$m_A = 1$	0, −50	0, 0
$m_A = 2$	0, −100	50, 0

$C_A = 1, C_B = 2$

	$m_B = 1$	$m_B = 2$
$m_A = 1$	−50, 0	−100, 0
$m_A = 2$	0, 0	0, 50

$C_A = 2, C_B = 1$

	$m_B = 1$	$m_B = 2$
$m_A = 1$	−50, −50	−100, 0
$m_A = 2$	0, −100	0, 0

$C_A = 2, C_B = 2$

Fig. 3.33 Game of incomplete information

3.4 Solution Concepts for Games in Normal Form

John C. Harsanyi, in a series of articles Harsanyi (1967), Harsanyi (1968a), Harsanyi (1968b), modeled games of incomplete information as *Bayesian games*, and proposed *Bayesian-Nash equilibrium* as a reasonable solution concept for such games.[14] In Bayesian games, nature[15] makes the first move and *privately signals* type $\theta_i \in \Theta_i$ to each Player i. Θ_i is the set of possible types of Player i. $\Theta = \Theta_1 \times \ldots \Theta_n$ is the set of type profiles where $\theta = (\theta_1, \ldots, \theta_n)$ is a type profile in the set Θ. There is a *unique common prior* F on the type profiles, and each player, knowing his own type, is intelligent enough to compute posterior probabilities on possible type profiles. Each type profile θ determines the state of the environment, and each state leads to a different game. Hence, posterior probabilities provide probabilities of which games are being played to each Player i, given his own type θ_i.

The game of incomplete information in Example 3.14 is a Bayesian game as the cost per dose is the type of each firm, which is private information. However, there is a unique common prior with each cost is equally likely. In this setting, the environment has four possible *states* (or type profiles)—$(C_A = 1, C_B = 1)$, $(C_A = 1, C_B = 2)$, $(C_A = 2, C_B = 1)$, $(C_A = 2, C_B = 2)$. If the cost per dose to firm A is one dollar ($C_A = 1$), then using the common prior, firm A can compute posterior probability as 0.5 for the two possible states $(C_A = 1, C_B = 1)$ and $(C_A = 1, C_B = 2)$. It is evident that when the state changes, the players' payoff changes.

In a Bayesian game $\langle N, (A_i), (u_i), \Theta, F \rangle$, a strategy function s_i of Player i is a mapping from the type set to the action set ($s_i : \Theta_i \mapsto A_i$). With this background, we can define Bayesian-Nash equilibrium.

Definition 3.11 (*Bayesian-Nash Equilibrium*) A Bayesian-Nash equilibrium for a Bayesian game $\langle N, (A_i), (u_i), \Theta, F \rangle$ is a strategy function profile $(s_1(\theta_1), \ldots, s_n(\theta_n))$, if for each θ_i and for each Player i

$$\sum_{(\theta_i, \theta_{-i})} F((\theta_i, \theta_{-i})|\theta_i) u_i(s_i(\theta_i), s_{-i}(\theta_{-i})) \geq \sum_{(\theta_i, \theta_{-i})} F((\theta_i, \theta_{-i})|\theta_i) u_i(s_i'(\theta_i), s_{-i}(\theta_{-i})) \quad \blacktriangleleft$$

θ_{-i} is the type profile of other players other than i. s_i' is another strategy function for Player i. $F((\theta_i, \theta_{-i})|\theta_i)$ is the posterior probability of state (θ_i, θ_{-i}), given Player i's type is θ_i. *Given the posterior probabilities for each possible state, unilateral deviations are suboptimal for each type of each player from the Bayesian-Nash strategy function profile.*

In both correlated equilibrium and Bayesian-Nash equilibrium, strategy functions are mapped to pure action sets. Using mixed strategies is complex as it involves randomization over the set of strategy functions. Furthermore, type set Θ_i can be infinite. First-price sealed-bid (FPSB) auction and second-price sealed-bid (SPSB) auction in Examples 3.4 and 3.10 are Bayesian games with valuation of each player can take any of the infinite values between 0 and 1.

---

[14] For this work, John C. Harsanyi shared the 1994 Nobel Memorial Prize in Economic Sciences with John Nash and Reinhard Selten.

[15] Recall footnote 5, nature is a non-strategic player.

52   3   Games in Normal Form

Let us compute Bayesian-Nash equilibrium of the vaccine procurement problem discussed in Example 3.14.

Example 3.15 (*Bayesian-Nash Equilibrium of Vaccine Procurement Problem*) The Bayesian game associated with the procurement mechanism is shown in Fig. 3.33. In any state (or game), if a firm's cost is 2, then messaging cost as 2 strictly dominates messaging 1 for that firm; and, if a firm's cost is 1, then messaging cost as 2 weakly dominates messaging 1 for that firm. For example, in the top red information partition when $C_A = 1$, Firm A *conjectures* that Firm B messages strictly dominant strategy in the game with $C_A = 1$, $C_B = 2$ and messages weakly dominant strategy in the game with $C_A = 1$, $C_B = 1$, then the best response of Firm A is to message $m_A = 2$. This reasoning is also applicable for Firm B when $C_B = 1$. Also, we have mentioned that when $C_A = 2$, messaging $m_A = 2$ is strictly dominant, and likewise for Firm B when $C_B = 2$. This means, in the Bayesian-Nash equilibrium of the mechanism, each firm messages cost as 2, irrespective of its true cost.   ◁

Problems

Problem 3.1 Compute the optimal strategy for each player in the matching pennies game (Example 3.2).   ◀

Problem 3.2 Compute pure and mixed strategy Nash equilibria in the following bimatrix game

Player 2

		L	R
	U	2, 1	0, 2
Player 1	**D**	1, 3	3, 0

◀

Problem 3.3 Consider the following bimatrix game.

Player 2

		L	R
	T	a, b	c, d
Player 1	**B**	e, f	g, h

i   If (T, L) is a dominant strategy equilibrium, then what inequalities must hold among $a, ..., h$?

ii If (T, L) is a Nash equilibrium, then what inequalities must hold among $a, ..., h$?

◀

Problems 53

Problem 3.4 Consider Cournot competition in a duopoly setting. Let firm i's profit be $u_i = q_i(\theta_i - q_i - q_j)$, where θ_i is the difference between the intercept of the linear demand curve and firm i's constant unit cost and q_i is the quantity chosen by firm i. It is common knowledge that $\theta_1 = 1$; while firm 1 believes that firm 2 has two equiprobable types $\theta_2^1 = \frac{5}{4}$ and $\theta_2^2 = \frac{3}{4}$. This belief is common knowledge. They choose their output simultaneously. Model this game as a Bayesian game and compute the unique BNE of the Bayesian game. ◀

Problem 3.5 Consider the network shown in Fig. 3.34. Six motorists wish to travel from Node 1 to Node 4. There are two routes $1 \to 2 \to 4$ and $1 \to 3 \to 4$. The motorists are rational and self-interested and can decide on the route to minimize travel time. The travel time for each link in minutes is shown in Fig. 3.34, where x_{ij} is the number of motorists using link ij. Compute Nash equilibria in this network.

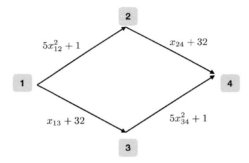

Fig. 3.34 Initial network

A link 23 is provided for instant commuting in both directions, as shown in Fig. 3.35. Compute Nash equilibria in the augmented network. ◀

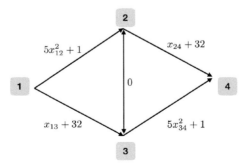

Fig. 3.35 Augmented network

Problem 3.6 Consider a supply chain with a manufacturer, M, and a retailer, R, for an electric scooter brand. The cost per unit c to the manufacturer is 11, and charges a price p per unit to the retailer. The market price P is determined by the retailer with market demand for scooters is $x = 100(131 - P)$. Customers can also buy from the manufacturer directly at price P. A customer's probability of buying from the retailer or the manufacturer is 0.5. The setting is shown in Fig. 3.36, with each channel equally likely. Assuming all the agents are self-interested and rational, compute p and P at Nash equilibrium.

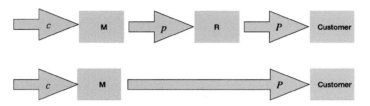

Fig. 3.36 Supply chain network ◀

Problem 3.7 (**Correlated Equilibrium using Traffic Signal**[16]) Two cars, A and B, are simultaneously approaching a junction, as shown in the figure below.

The drivers can choose between "Go" and "Stop". If both choose "Go", it leads to an accident with a high negative payoff. The payoff matrix is

		Car B Go	Car B Stop
Car A	Go	−10, 10	5, 0
	Stop	0, 5	−1, −1

To coordinate their movement at the junction, a traffic signal is installed that acts like a coordination device. The traffic signal recommends green (for "Go") or red

---

[16] Adapted from Owen (2001).

(for "Stop") to each driver. Assume no policeman or other devices to enforce the recommendations, compute correlated equilibria of this setting.[17] ◀

References

Aumann, R. J. (1974). Subjectivity and correlation in randomized strategies. *Journal of Mathematical Economics, 1*(1), 67–96.

Braess, D. (1968). Über ein paradoxon aus der verkehrsplanung. *Unternehmensforschung, 12*(1), 258–268.

Chen, X., Deng, X., & Teng, S. H. (2009). Settling the complexity of computing two-player Nash equilibria. *Journal of the ACM (JACM), 56*(3).

Chwe, M.S.-Y. (2013). *Rational ritual: Culture, coordination, and common knowledge.* Princeton University Press.

Ferguson, T. S. (2020). *A course in game theory.* World Scientific Publishing Co Pte Ltd.

Harsanyi, J. C. (1967). Games with incomplete information played by Bayesian players, I-III Part I: The basic model. *Management Science, 14*(3), 159–182.

Harsanyi, J. C. (1968a). Games with incomplete information played by bayesian players, Part III. The basic probability distribution of the game. *Management Science, 14*(7), 486–502.

Harsanyi, J. C. (1968b). Games with incomplete information played by bayesian players part II. Bayesian Equilibrium Points. *Management Science, 14*(5), 320–334.

Harsanyi, J. C., & Selten, R. (1988). *A general theory of equilibrium selection in games.* MIT Press.

Kakutani, S. (1941). A generalization of brouwer's fixed point theorem. *Duke Mathematical Journal, 8*(3), 457–459.

Kreps, D. M. (1990). *A course in microeconomic theory.* Princeton, N.J.: Princeton University Press.

Lemke, C. E., & Howson, J. T. (1964). Equilibrium points of bimatrix games. *Journal of the Society for Industrial and Applied Mathematics, 12*(2).

Lewis, D. (2002). *Convention: A philosophical study.* Blackwell.

Luenberger, D. G. & Ye, Y. (2008). Linear and nonlinear programming. In *International Series in Operations Research & Management Science* (Vol. 16). US, New York, NY: Springer.

Nash, J. (1951). Non-cooperative games. *The Annals of Mathematics, 54*(2), 286.

Osborne, M. J., & Rubinstein, A. (1994). *A course in game theory.* Cambridge, Mass: MIT Press.

Owen, G. (2001). *Game theory.* San Diego, Calif: Academic.

Pugh, C. C. (2002). *Real mathematical analysis.* New York: Springer.

Rawls, J. (1971). *A theory of justice.* Belknap Press of Harvard University Press.

Roughgarden, T. (2016). *Twenty lectures on algorithmic game theory.* Cambridge University Press.

Shapley, L. S. (1974). *A note on the Lemke-Howson algorithm. Part of the mathematical programming studies book series* (MATHPROGRAMM, Vol. 1, pp. 175–189).

Shoham, Y., & Leyton-Brown, K. (2009). *Multiagent systems- algorithmic.* Game Theoretic and Logic Foundation: Cambridge University Press.

Vickrey, W. (1961). Counterspeculation, auctions, and competitive sealed tenders. *The Journal of Finance, 16*(1), 8–37.

von Neumann, J. (1928). Zur Theorie der Gesellschaftsspiele. *Mathematische Annalen, 100*(1), 295–320.

von Stengel, B. (2021). Finding Nash Equilibria of Two-Player Games. arXiv:2102.04580.

von Stengel, B. (2022). Zero-Sum Games and Linear Programming Duality. arXiv:2205.11196.

---

[17] Hint: compute the probabilities with which each outcome of the bimatrix game can be selected by the traffic signal so that unilateral deviation from the recommendations is suboptimal for each driver.

Chapter 4
Games in Normal Form: Applications in OM

In Chap. 3, we discussed the theoretical underpinnings of games in normal form. This chapter provides a concise overview of important operations management (OM) applications[1] that can be represented in normal-form representation.

4.1 Inventory Games[2]

Newsvendor models effectively capture the fundamental principles of operations management. In its most basic iteration, as elucidated in Chap. 1, a decision-maker determines the quantity by carefully weighing the costs associated with inadequate stock levels against the expenses incurred from excessive stock levels. Numerous applications for such models span across fields characterized by transient and uncertain demands. The fundamental assumption is that the decision-maker possesses exclusive authority to influence her payoffs. However, it is important to acknowledge that, apart from uncertainty in demand, the environment can have multiple decision-makers, each driven by self-interest and aiming to control the environment to improve their payoffs. Under these conditions, applying game theory to newsvendor models becomes pertinent and crucial.

The idea of the competitive newsvendor emanated from a single decision-maker selling two or more products. Mcgillivray and Silver (1978) examine the issue of product replacement within the economic order quantity (EOQ) framework, whereas Parlar and Goyal (1984) investigate the same topic within the classical newsvendor setting, and introduce a game-theoretic approach to the newsvendor problem, which

---

[1] It is important to note that the notations used in each section in this chapter are specific to that application.

[2] We focus on inventory decisions under uncertain demand using the newsvendor framework. In this section, the OM actions are on the *asset side* in the VCAP framework (refer Sect. 2.3).

© The Author(s), under exclusive license to Springer Nature Singapore Pte Ltd. 2024 57
R. K. Amit, *Game Theory with Applications in Operations Management*, Springer Texts in Business and Economics, https://doi.org/10.1007/978-981-99-4833-8_4

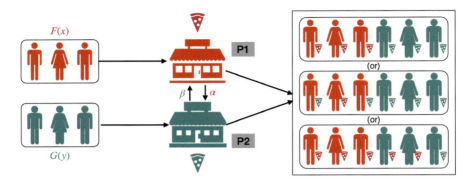

Fig. 4.1 Newsvendor game

is widely recognized as *newsvendor games*. The fundamental premise that renders game theory the right analytical tool is the assumption that, within this context, buyers can purchase from an alternative newsvendor if their chosen newsvendor runs out of stock. Consequently, the profitability of a newsvendor is contingent upon their quantity decisions as well as the judgments made by other newsvendors.

For illustration, we present a model, results, and insights based on Parlar's work (Parlar, 1988). We consider two decision-makers, P1 and P2, with order quantities m and n. To simplify notation, we use the index $i \in \{1, 2\}$ to represent the parameters of these decision-makers. Specifically, let s_i, c_i, p_i, and q_i be the order cost, selling price, backorder cost, and salvage value per unit for player i's product. The demand for P1 and P2 is distributed with probability density functions $f(\cdot)$ and $g(\cdot)$, respectively, and cumulative distribution functions $F(x) = \int_0^x f(t)\, dt$ and $G(y) = \int_0^y g(t)\, dt$. Furthermore, we define $\alpha \in [0, 1]$ and $\beta \in [0, 1]$ as the fractions of demand from P1 and P2, respectively, that will switch to P2 and P1 when P1 and P2 are stocked out. The setting is illustrated in Fig. 4.1.

Along with these definitions, the following assumptions are similar to Cournot conjecture in Cournot duopoly (Example 3.3) to derive meaningful insights. Firstly, the shelf life of products is one period, and the game starts with zero units. Secondly, substitution rates are applicable and will occur if the preferred newsvendor runs out of stock. Thirdly, each player is assumed to be rational, i.e., a player will not risk his payoffs to reduce the other player's payoffs. Finally, the typical assumptions like $s_i > c_i$ and $c_i > q_i \geq 0, i = \{1, 2\}$ are made. It's important to note that although we present the analysis primarily from P1's perspective for the sake of clarity, a similar type of analysis applies to P2 as well. Given these assumptions, depending on the values the random variables take $X = x$ and $Y = y$, P1's expected profit $J_1(m, n)$ is given by

4.1 Inventory Games

$$J_1(m, n) = \int_0^m \int_0^n [s_1 x + q_1(m - x) - c_1 m] f(x)g(y) \, dy \, dx \tag{4.1}$$

$$+ \int_0^m \int_n^\infty \{s_1 x + s_1 \min[\beta(y - n), m - x]$$

$$+ q_1 \max[0, (m - x) - b(y - n)] - c_1 m\} f(x)g(y) \, dy \, dx$$

$$+ \int_m^\infty \int_0^n [s_1 m - p_1(x - m) - c_1 m] f(x)g(y) \, dy \, dx$$

$$+ \int_m^\infty \int_n^\infty [s_1 m - p_1(x - m) - c_1 m] f(x)g(y) \, dy \, dx$$

where first, second, third, and fourth terms represent the situations $\{x \leq m, y \leq n\}$, $\{x \leq m, y \geq n\}$, $\{x \geq m, y \leq n\}$, and $\{x \geq m, y \geq n\}$, respectively. Using Definition 2.1, the normal-form representation $\langle N, a_i, u_i \rangle$ with $N = \{P1, P2\}$, $a_i \in [0, \infty)$ and $u_i = J_i(m, n)$ is a newsvendor game.

In reference with Definition 3.9, a pair (m^*, n^*) constitutes Nash equilibrium if

$$J_1(m^*, n^*) \geq J_1(m, n^*) \; \& \; J_2(m^*, n^*) \geq J_2(m^*, n) \tag{4.2}$$

Equilibrium is computed by the intersection of the reaction curves of both players. P1's reaction curve is the collection of points $\{(m, n) \mid J_1(m, n) \geq J_1(m', v), m' \geq 0\}$ and found by equating first differential with respect to m to 0 conditioned on J_1 being continuously differentiable and concave with respect to m for each n. After some simplifications and through Leibnitz's differentiation rule, the first differential of J_1 with respect to m is

$$\frac{\partial J_1}{\partial m} = (s_1 + p_1) \int_u^\infty f(x) \, dx + q_1 \int_0^u f(x) \, dx + (s_1 - q_1)$$

$$\int_0^u \int_B^\infty g(y)f(x) \, dy \, dx - c_1 = 0 \tag{4.3}$$

where $B = [(m - x)/b] + n$. The second partial indicates that $\partial^2 J_1/\partial m^2 < 0$, i.e., P1's expected profit function is strictly concave in m for each n. We need two more results to prove the existence and uniqueness of equilibrium. For the considered newsvendor game, the upper and lower bounds of the action set for any player can be obtained in Eq. 4.3 by considering the other player's decision variable to be 0 and ∞, respectively. The intuition behind the numbers used for bounds is straightforward. Specifically, if $\overline{m}$ and $\underline{m}$ are the upper and lower bounds for player P1, then

$$\overline{m} = \{m : \partial J_1/\partial m = 0 \text{ when } n = 0\} \tag{4.4}$$

$$\underline{m} = \{m : \partial J_1/\partial m = 0 \text{ when } n \to \infty\} \tag{4.5}$$

As $\partial J_1/\partial m = 0$ is a curve in (m, n) plane, monotonicity properties are necessary for proving the uniqueness of the equilibrium. Lemma 4.1, adapted from Parlar (1988), aids us in this regard. Let $I_1 = \partial J_1/\partial m$.

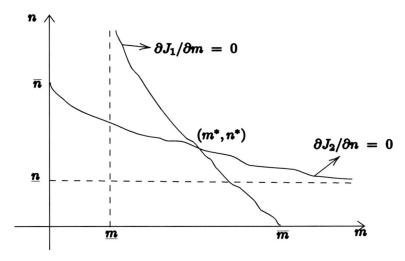

Fig. 4.2 Reaction curves (Adapted from Parlar, 1988)

Lemma 4.1 (Parlar, 1988) *$I_1(m, n) = 0$ is a strictly decreasing curve in the (m, n) plane.* ◀

It is easy to prove Lemma 4.1 by differentiating I_1 with respect to n that yields $\partial I_1/\partial n < 0$, i.e., Eq. 4.3 is strictly decreasing curve in (m, n) plane. As mentioned earlier, parallel arguments for P2 yield similar results, and the reaction curves for both players are shown in Fig. 4.2. With these results, the uniqueness of the equilibrium can be easily established.

Theorem 4.1 (Parlar, 1988) *There exists a unique Nash solution (m^*, n^*).* ◀

As the sufficiency conditions for proving Theorem 4.1 are more involved, interested readers can refer to Parlar (1988) for better understanding. However, the intuition is clear that because of the boundedness and monotone properties of the reaction curves, the equilibrium with Nash property exists and is unique, which can also be verified from Fig. 4.2.

To bring more clarity to the analysis, we provide a numerical example borrowed from Wu and Parlar (2011). Let the demand densities be exponential, i.e., $f(x) = \lambda e^{-\lambda x}$ and $g(y) = \mu e^{-\mu y}$ where $\lambda = 1/30$ and $\mu = 1/20$. And, the values of other parameters are $s_1 = 15$, $s_2 = 9$, $c_1 = 8$, $c_2 = 5$, $p_1 = p_2 = q_1 = q_2 = 0$, and $\alpha = \beta = 0.9$. For this setting, $m^* = 25.38$ and $n^* = 19.55$, with expected profits as $J_1 = 83.63$ and $J_2 = 35.91$.

However, yet impossible, for comparison, if we allow the players to choose their and the competitor's quantity, the results are as follows: when P1 chooses freely, $m = 37.21$, $n = 0$ and $J_1 = 148.11$; when P2 is allowed to choose freely, $m = 0$, $n = 35.21$ and $J_2 = 80.91$. A quick look at the numbers indicates that the expected profits of players decrease when we consider competition and the industry inventory increases (44.93 vs. 37.21 or 35.21). This inference is true in most real-world applications, and we would like to quote a recent example from the Indian context. Aggregate inventory is one of the important components of National Accounts. In India, inventory increased by ₹447.66 Billion in the fourth quarter of 2021 (Source: https://tradingeconomics.com). In such instances, game theory guides understanding and explaining the issues in inventory management.

The research paradigm on inventory games has come a long way since (Parlar, 1988), the fundamental insight remains the same. One much-needed and realistic extension is to relax the complete information approach, i.e., players may not have complete information of other players' backorder cost, salvage value, and demand. Furthermore, the exploration of coordination mechanisms, such as contracts (see Sect. 4.4) or collaboration (see Sect. 8.1), is also a viable avenue to address the inefficiencies resulting from non-cooperation.

4.2 Traffic Planning[3]

In Example 3.12, we discussed Braess's paradox, wherein the counter-intuitive phenomenon of increased travel time resulting from the provision of an extra path intended to alleviate traffic congestion was demonstrated. To demonstrate the significance, we provide two historical examples. According to Knödel (2013), the City of Stuttgart constructed a new roadway to mitigate traffic congestion. However, the impact on traffic flow was negative, ultimately resulting in the street's closure. An additional illustration pertains to the inverse manifestation of Braess's conundrum. In New York City on Earth Day in 1990, the Transportation Commissioner ordered to close 42nd Street. Kolata (1990) observes that traffic congestion unexpectedly decreased, leading to an improvement in the overall traffic flow.

One potential criticism of Braess's paradox is that the contradiction arose due to the game's inherent construction, specifically its reliance on numerical elements. This section elucidates the concept by examining a comprehensive network framework. Our discussion closely follows Pas and Principio (1997).

Consider a simple, symmetric network shown in Fig. 4.3 with linear travel times. Suppose the total demand from origin to destination is Q and let $\alpha_{ij} \geq 0$ denote the free flow travel time and $\beta_{ij} > 0$ denote the delay parameter due to the flow $f_{ij} \geq 0$. Then, the total travel time on link (i, j) is

$$t_{ij} = \alpha_{ij} + \beta_{ij} f_{ij} \qquad (4.6)$$

Because of symmetry in the network, $\alpha_{ab} = \alpha_{cd}, \alpha_{ac} = \alpha_{bd}$, and $\beta_{ab} = \beta_{cd}, \beta_{ac} = \beta_{bd}$. Since the links ab and cd are very short, they are assumed to have null free flow time, i.e., $\alpha_{ab} = \alpha_{cd} = 0$. Finally, for the fifth link, assume $\beta_{bc} = \beta_{ac} = \beta_{bd}$. For notational convenience, let

---

[3] In this section, the OM actions are on the *process side* in the VCAP framework.

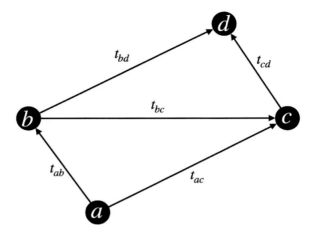

Fig. 4.3 Four node network

$$\alpha_{ac} = \alpha_{bd} = \alpha_1,$$
$$\alpha_{bc} = \alpha_2,$$
$$\beta_{ab} = \beta_{cd} = \beta_1,$$
$$\beta_{bc} = \beta_{ac} = \beta_{bd} = \beta_2$$

Given this rephrased travel time parameters, the updated total travel times are

$$t_{ab} = \beta_1 f_{ab}$$
$$t_{cd} = \beta_1 f_{cd}$$
$$t_{ac} = \alpha_1 + \beta_2 f_{ac}$$
$$t_{bd} = \alpha_1 + \beta_2 f_{bd}$$
$$t_{bc} = \alpha_2 + \beta_2 f_{bc}$$

The three potential paths to reach the destination d from the source a are $a \to b \to d$, $a \to c \to d$, and $a \to b \to c \to d$. The total travel time in a path is the sum of the travel times of the links in the corresponding path. For instance, total travel time through the path $a \to b \to d$, $t^{abd} = t_{ab} + t_{bd}$, and its flow is denoted by f^{abd}. As a result, $Q = f^{abd} + f^{acd} + f^{abcd}$.

Schulz and Stier Moses (2002) establish that the user-optimized equilibrium always exists and is unique in a network without capacities. Moreover, in a symmetric network, flows and travel times in all the potential paths are equal.

In the four-link network (*without* the link bc), equilibrium flows are $f^{abd} = f^{acd} = Q/2$, and the respective travel times are

4.2 Traffic Planning

$$t^{abd} = t^{acd} = \beta_1 f_{ab} + \alpha_1 + \beta_2 f_{bd} = \beta_1 f_{cd} + \alpha_1 + \beta_2 f_{ac}$$
$$= \frac{Q\beta_1}{2} + \alpha_1 + \frac{Q\beta_2}{2} = \frac{Q(\beta_1 + \beta_2)}{2} + \alpha_1$$

Given this user equilibrium, the total system travel time T^4 is given by

$$T^4 = t^{abd} f^{abd} + t^{acd} f^{acd} = \frac{Q}{2} t^{abd} + \frac{Q}{2} t^{acd} = \frac{Q^2(\beta_1 + \beta_2)}{2} + Q\alpha_1 \quad (4.7)$$

Similarly, equilibrium in the five-link network (*with* link bc) can be obtained by equating the travel times on all three possible paths, $t^{abd} = t^{acd} = t^{abcd}$. Here, the flows are as follows:

$$f^{abd} = f^{acd} = \frac{\alpha_2 - \alpha_1 + Q(\beta_1 + \beta_2)}{\beta_1 + 3\beta_2} \quad (4.8)$$

$$f^{abcd} = Q - 2f^{abd} \quad (4.9)$$

and

$$t^{abd} = t^{acd} = t^{abcd} = \alpha_1 + Q\beta_1 + (\beta_2 - \beta_1) \left[\frac{\alpha_2 - \alpha_1 + Q(\beta_1 + \beta_2)}{\beta_1 + 3\beta_2} \right] \quad (4.10)$$

In Eqs. 4.8–4.10, if $Q \leq \frac{\alpha_1 - \alpha_2}{\beta_1 + \beta_2}$, all the traffic flows the route $a \rightarrow b \rightarrow c \rightarrow d$. It means $f^{abd} = f^{acd} = 0$ and $f^{abcd} = Q$, which leads to

$$t^{abcd} = Q(2\beta_1 + \beta_2) + \alpha_2 \quad (4.11)$$

Also, in Eqs. 4.8–4.10, if $Q \geq \frac{2(\alpha_1 - \alpha_2)}{\beta_1 - \beta_2}$, then $a \rightarrow b \rightarrow c \rightarrow d$ carries zero traffic, whereas $a \rightarrow b \rightarrow d$ and $a \rightarrow c \rightarrow d$ equally share the total flow Q. As a result,

$$t^{abd} = t^{acd} = \frac{Q(\beta_1 + \beta_2)}{2} + \alpha_1 \quad (4.12)$$

Finally, for the values of Q between $\frac{\alpha_1 - \alpha_2}{\beta_1 + \beta_2}$ and $\frac{2(\alpha_1 - \alpha_2)}{\beta_1 - \beta_2}$, all the routes carry equal amount of traffic and its travel time is similar to Eq. 4.10. Given these individual travel times, the total system travel time for the five-link network is

$$T^5 = \begin{cases} Q^2(2\beta_1 + \beta_2) + Q\alpha_2, & Q \leq \dfrac{\alpha_1 - \alpha_2}{\beta_1 + \beta_2} \\[3mm] Q\left(\alpha_1 + Q\beta_1 + (\beta_2 - \beta_1)\left[\dfrac{\alpha_2 - \alpha_1 + Q(\beta_1 + \beta_2)}{\beta_1 + 3\beta_2}\right]\right), & \dfrac{\alpha_1 - \alpha_2}{\beta_1 + \beta_2} \leq Q \leq \dfrac{2(\alpha_1 - \alpha_2)}{\beta_1 - \beta_2} \\[3mm] \dfrac{Q^2(\beta_1 + \beta_2)}{2} + Q\alpha_1, & Q \geq \dfrac{2(\alpha_1 - \alpha_2)}{\beta_1 - \beta_2} \end{cases}$$
$$(4.13)$$

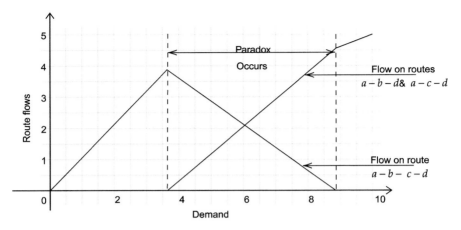

Fig. 4.4 Regions of paradox (Adapted from Pas & Principio, 1997)

The necessary conditions for Braess's paradox to occur ($T^5 > T^4$) are

$$T^5 > T^4 \text{ iff } Q(\beta_1 - \beta_2) < 2(\alpha_1 - \alpha_2) \text{ or } Q(3\beta_1 + \beta_2) > 2(\alpha_1 - \alpha_2) \quad (4.14)$$

equivalently,

$$T^5 > T^4 \text{ iff } \frac{2(\alpha_1 - \alpha_2)}{(3\beta_1 + \beta_2)} < Q < \frac{2(\alpha_1 - \alpha_2)}{\beta_1 - \beta_2} \quad (4.15)$$

Equation 4.15 yields several interesting insights. Firstly, the paradox will not occur for any value of Q if the free flow travel times in the network are equal, $\alpha_1 = \alpha_2$. Secondly, the upper limit of the range is finite unless $\beta_1 = \beta_2$. Lastly, the surprising and counter-intuitive result is that Braess's paradox will not occur for "too large" or "too small" travel demands. These insights reason out the two real-life paradoxes highlighted at the start of this section. Moreover, to clarify it further, for the parameter values $\alpha_1 = 50$, $\alpha_2 = 10$, $\beta_1 = 10$, $\beta_2 = 1$, Fig. 4.4 shows the range of the demand where the paradox occurs.

The paradox, which was initially introduced to the English community by Murchland (1970), has stimulated scholars from several fields and has become an integral component of game theory. Applications can be observed in several disciplines, such as computer science, electrical engineering, physics, biology, and sports. We direct readers to Nagurney and Nagurney (2021) to explain the described applications comprehensively.

4.3 Airline Alliances[4]

Have you ever wondered how a ticket procured from a singular airline within a multi-leg schedule encompasses a flight executed by a distinct airline? Have you ever pondered the mechanism by which two distinct airlines, perhaps in a state of rivalry, have provided you with a consolidated ticket to facilitate your entire travel itinerary?

Let us contemplate the scenario of a traveler embarking on a journey from Chennai, India to Melbourne, Australia. Among the several alternatives at one's disposal, a particular option is presented in Fig. 4.5. It is important to observe that Air India conducts flights on both segments of the above itinerary. Yet, they share the capacity with Qantas Airways.

Two key inquiries pertinent to our discourse are: (i) what are the reasons behind airlines' inclination to share capacity with other airlines, potentially even with competitors? (ii) What is its competitive advantage, and how might the decisions be altered to mitigate the risk of losing market share?

An alliance refers to the formation of a group consisting of two or more parties to attain mutual benefits. These strategic alliances confer a competitive edge to the participating entities and prove beneficial in accessing markets where the entity lacks a fleet presence. According to Van Mieghem and Allon (2015), partnerships can be categorized within the framework of traditional operations management as either subcontracting or outsourcing approaches, according to the level of flexibility they offer. Remarkably, alliances are prevalent within the maritime and liner transport

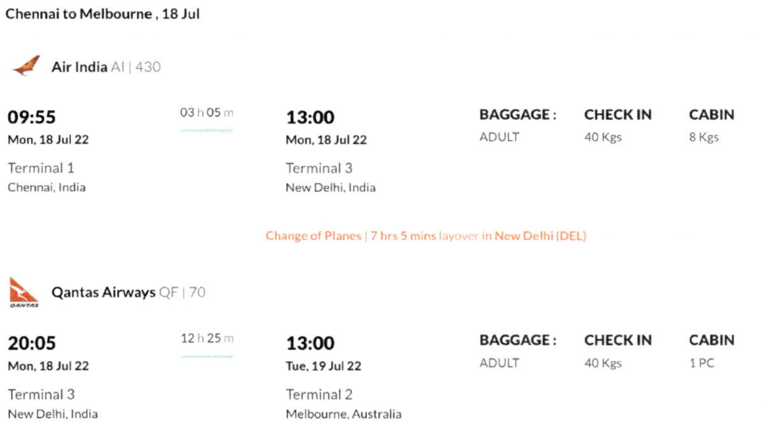

Chennai to Melbourne , 18 Jul

Air India AI | 430

09:55	03 h 05 m	**13:00**		**BAGGAGE :**	**CHECK IN**	**CABIN**
Mon, 18 Jul 22		Mon, 18 Jul 22		ADULT	40 Kgs	8 Kgs
Terminal 1		Terminal 3				
Chennai, India		New Delhi, India				

Change of Planes | 7 hrs 5 mins layover in New Delhi (DEL)

Qantas Airways QF | 70

20:05	12 h 25 m	**13:00**		**BAGGAGE :**	**CHECK IN**	**CABIN**
Mon, 18 Jul 22		Tue, 19 Jul 22		ADULT	40 Kgs	1 PC
Terminal 3		Terminal 2				
New Delhi, India		Melbourne, Australia				

Fig. 4.5 Capacity sharing by Air India and Qantas Airways

---

[4] In this section, the OM actions are on the *process side* in the VCAP framework.

sectors, with shipping alliances exhibiting substantial scale in cargo volume and participating entities.

As an illustration, it is noteworthy that a significant proportion of over 50 percent of maritime freight transportation is facilitated by two dominant entities, namely, 2M and Ocean Alliance. While the size of shipping alliances may surpass them, it is worth noting that the three prominent alliances, namely, Star Alliance, SkyTeam, and Oneworld, collectively account for around 45% of worldwide traffic (Pearson, 2022). As a result of these strategic partnerships, airlines have the ability to collaborate on capacity utilization and market seats operated by a separate carrier, a practice sometimes referred to as *code sharing*. Code sharing refers to the practice in the airline industry wherein an airline engages in marketing and ticket sales for flights operated by another airline. A type of airline that engages in the former practice is commonly referred to as a marketing carrier, while the latter practice is commonly associated with an operating carrier.

Figure 4.6 depicts the concept of code sharing followed by two carriers AA and BB. Since AA does not operate its flight from Location B to C, it enters into a codeshare agreement with BB, leading to few AA passengers traveling on BB flight in segment B-C. While codeshare arrangements and airline alliances encompass a wide spectrum of studies, we focus on a basic codeshare agreement to demonstrate the potential applications of game theory.

In the context of codeshare arrangements, it is customary to employ cooperative game theory as a modeling framework. This approach entails airlines negotiating to determine the distribution of money derived from connecting passengers. After completion, each airline uses its independent discretion to determine the *booking limits* based on the agreement and then communicates the updated real-time seat

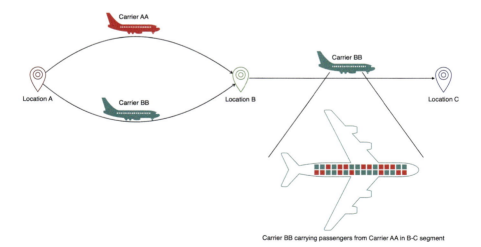

Fig. 4.6 Codesharing

4.3 Airline Alliances

availability to its partner. Hu et al. (2013) provide an excellent commentary on the issues in implementing airline alliances.

It is crucial to comprehend the notion of booking limit and its significance in the field of airline revenue management. Generally, the demand for airline services encompasses high-value and low-value consumer segments, with empirical evidence indicating that low-value customers tend to arrive before high-value customers. Despite receiving access to the same product, customers from various groups are charged varying fares, known as fare classes, based on their willingness to pay. To enhance the cost-effectiveness of their services, clients in higher fare classes are provided with supplementary advantages such as more lenient limits on travel rescheduling and exemption from cancellation fees. In light of this particular demand, the airline faces a trade-off whereby it must decide between catering to consumers with lower value and those with higher value. One potential approach entails safeguarding the seats designated for high-value consumers by implementing booking limits, effectively restricting the seats available in the low-fare class. For an introduction to booking limits with two-fare classes, the interested reader might consult Littlewood (1972). Similarly, Belobaba (1989) provides an introduction to booking limits with multiple fare classes.

In this section, we use noncooperative games to determine the optimal booking limits for the airline, taking into account factors such as local and connecting traffic demands, revenue considerations, and the existing alliance agreement. Let us consider a simple alliance between two airlines (Airline 1 and Airline 2) that operate flights on successive legs depicted in Fig. 4.7. Airlines 1 and 2 provide services on two distinct flight routes: A-B and B-C, respectively. These routes offer three potential travel itineraries: two local itineraries, specifically A-B and B-C, and a connecting itinerary that spans A-B-C. According to Netessine and Shumsky (2005), the structure described can be classified as "vertical competition", wherein airlines operate in distinct markets that do not overlap, but yet compete with each other through the transportation of connecting passengers. Moreover, this particular framework exemplifies a common instance of competition, when airlines interact with one another to establish connections with passengers and operate autonomously while also serving the needs of local clientele.

It is assumed that the revenue generated by a connecting passenger (itinerary A-B-C) is equal to the sum of the local itineraries, and the codeshare agreement is such that the revenue split is equal to the revenue from the local itinerary operated by each airline. Let $j = \{1(A, B), 2(B, C)\}$ and $k = \{L, H\}$ denote the legs, and fare

Fig. 4.7 Airline alliance

68 4 Games in Normal Form: Applications in OM

Table 4.1 Demand parameters

Notation	Description
D_{Lj}	Local low-fare demand specific to leg j
D_{Hj}	Local high-fare demand specific to leg j
$\hat{D}_{Lj}(\overline{B}_{-j})$	Connecting low-fare demand specific to leg j, given the decision on $-j$
$\hat{D}_{Hj}(\mathbf{B})$	Connecting high-fare demand specific to leg j, given the decision on $-j$

classes, respectively. As each airline operates on a particular, airlines and legs can be used interchangeably.

Table 4.1 provides the demand details specific to leg j. $\mathbf{B} = (\overline{B}_1, \overline{B}_2)$ denote the two-dimensional vector where the first dimension is the booking limit of Airline 1 on leg 1(A, B), and the second dimension is the booking limit of Airline 2 on leg 2(B, C). Additionally, we assume that $\hat{D}_{Lj}(\overline{B}_{-j})$ is increasing in $\overline{B}_{-j}$ and $\hat{D}_{Hj}(\mathbf{B})$ is decreasing in $\mathbf{B}$ as a higher booking limit in one leg provides a higher number of seats for connecting low-fare customers for the other airline.

With this background, we define *airline alliance game* as a noncooperative game with normal-form representation. In the game, the airline operating on leg j determines its booking limit $B_j \in [0, C_j]$ to maximize its expected profit π_j, assuming the airline operating on the other leg is acting similarly. C_j denotes the number of seats in the airline operating on leg j. For the airline operating on leg j, the expected profit π_j for the considered setting is given by

$$\begin{aligned} \pi_j = &\mathbb{E}[p_{Lj} \min(D_{Lj} + \hat{D}_{Lj}(\overline{B}_{-j}), B_j) \\ &+ p_{Hj} \min(D_{Hj} + \hat{D}_{Hj}(\mathbf{B}), C_j - \min(D_{Lj} + \hat{D}_{Lj}(\overline{B}_{-j}), B_j)] \end{aligned} \tag{4.16}$$

For benchmarking, Netessine and Shumsky (2005) consider a monopolistic setting where both the legs are controlled by a single airline. Proposition 4.1 sheds light on the comparison by simplifying their result modified to the considered setting.

Proposition 4.1 (Netessine & Shumsky, 2005) *If π_j is supermodular in $\overline{B}_j$ and $\partial \pi_j / \partial B_{-j} \geq 0$, then $\overline{\mathbf{B}}^c \leq \overline{\mathbf{B}}^m$. If π_j is supermodular in $\overline{B}_j$ and $\partial \pi_i / \partial B_{-i} \leq 0$, then $\overline{\mathbf{B}}^c \geq \overline{\mathbf{B}}^m$. $\overline{\mathbf{B}}^c$ and $\overline{\mathbf{B}}^m$ denote the equilibrium booking limit and the monopoly's booking limit.* ◀

Breaking down the problem into two extremes helps interpret the results of Proposition 4.1. Consider all the connecting passengers are in the low fare. Then, Eq. 4.16 reduces to

$$\begin{aligned} \pi_j = \mathbb{E}[p_{Lj} \min(D_{Lj} + \hat{D}_{Lj}(\overline{B}_{-j}), B_j) + p_{Hj} \min(D_{Hj}, C_j - \min(D_{Lj} \\ + \hat{D}_{Lj}(\overline{B}_{-j}), B_j)] \end{aligned} \tag{4.17}$$

4.3 Airline Alliances

Assuming that the distribution of (D_{Li}, D_{Hi}) is TP_2,[5] it can be verified that $\partial \pi_j / \partial B_{-j} \geq 0 \implies \overline{\mathbf{B}}^c \leq \overline{\mathbf{B}}^m$ as $\hat{D}_{Lj}(\overline{B}_{-i})$ is increasing in $\overline{B}_{-j}$. In simple terms, when the airline expects only low-fare connecting passengers, it prefers to set its booking limit lower when compared with the network optimality. At the other extreme, where all the connecting passengers are of high value, the modified expected profit function will be

$$\pi_j = \mathbb{E}[p_{Lj} \min(D_{Lj}, B_j) + p_{Hj} \min(D_{Hj} + \hat{D}_{Hj}(\overline{\mathbf{B}}), C_j - \min(D_{Lj}, B_j))] \tag{4.18}$$

Similar to the former extreme, as $\hat{D}_{Hj}(\overline{\mathbf{B}})$ is decreasing in $\mathbf{B}$, $\overline{\mathbf{B}}^c \leq \overline{\mathbf{B}}^m$, i.e., airline prefers to set higher booking limits than network optimality if the connecting passengers are expected to be of high value. With these two extremes, an analysis of intermediate cases where the connecting demand is a mix of high-value and low-value traffic can be derived. Results indicate that the difference in booking limits of competitive and monopoly increases with the increase in connecting traffic and the fraction of high-value customers in the connecting traffic. Summarizing the implications, the game-theoretic lens helps in debriefing how airlines should behave in the presence of codeshare agreements.

One recurring topic in the analysis is the observation that competition among airlines results in unsatisfactory outcomes from a social perspective. Specifically, the booking restrictions set by airlines tend to deviate from the optimal network capacity, either exceeding or falling short of the ideal level. Therefore, it is evident that the aggregate revenues in this competitive context will consistently be no superior to those in the centralized setting. The phenomenon known as *double marginalization*, which refers to the situation when local optimization at the level of each firm or level results in a global suboptimal solution, was first identified in Spengler (1950). In order to address this inefficiency, the implementation of contracts can be employed.

A comprehensive examination of contracts from a game-theoretic perspective is provided in Sect. 4.4. In the airline industry, game theory finds application in the form of airline alliances. Nevertheless, considering the inherent characteristics of the airline sector, characterized by narrow profit margins and intense competition, the years to come would necessitate the use of complex game-theoretic models.

---

[5] Totally positive of order 2, TP_2 implies that the realizations of random variables are more likely to be positively correlated than negatively correlated. Refer Joe (1997) for a deeper understanding of the concept.

4.4 Supply Chain Contracts[6]

During the early twentieth century, the manufacturing sectors operated under a system of *vertical integration*, when a single enterprise assumed responsibility for all aspects of production, ranging from the procurement of raw materials to the identification of suitable markets for the distribution of finished goods. One example of vertical integration is Ford Motors throughout the 1930s. High transaction costs rationalize vertical integration (Coase, 1937).[7] The emergence of technology has led to a significant reduction in transaction costs, which shrinks the boundaries of enterprises (Williamson, 1995).[8]

Figure 4.8 shows the boundary of vertically integrated Ford Motors in the 1930s. With advances in technology, the boundaries of enterprises started shrinking, as shown for Toyota Motors in the dotted line in Fig. 4.8. Some of the relevant technologies that enable the transition are shown in Fig. 4.9.

With shrinking boundaries, many business activities are getting outsourced, which necessitates the establishment of connections between various entities to facilitate the movement of finished goods from the raw material stage to the market. This interconnected network of entities is commonly referred to as a *supply chain*. As mentioned in Chap. 1, these developments resulted in scenarios of conflict and coop-

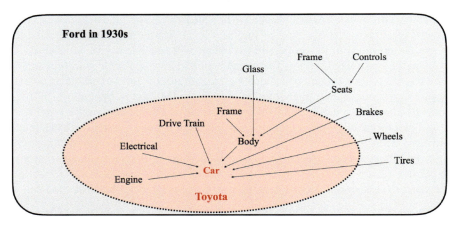

Fig. 4.8 Boundaries of enterprise (Adapted from Kreps, 2004)

---

[6] In this section, the OM actions are on the *process side* in the VCAP framework.

[7] Ronald Coase and Oliver Williamson were, respectively, awarded 1991 and 2009 Nobel Memorial Prizes in Economic Sciences for their work on transaction cost economics.

[8] A recent article in *The Economist* titled "How technology is redrawing the boundaries of the firm" provides a discussion on how digital technologies are allowing companies to reorganize themselves (The Economist; Jan 8, 2023).

4.4 Supply Chain Contracts

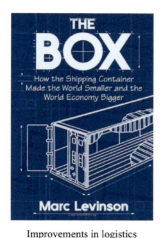

IBM's MAPICS and COPICS
for Material Requirements Planning (MRP)

Enterprise Resource Planning (ERP)

Visibility in production improves

Operations Research Models

Improvements in logistics

Fig. 4.9 Technologies that impact transaction costs

eration among the supply chain partners,[9] and a supply chain needs to be coordinated to create overall value. One of the key inhibitors of coordination is *double marginalization* in supply chains. Scholarly investigations into supply chain coordination via contractual arrangements commenced during the mid-1990s, despite double marginalization being established as early as Spengler's work in 1950. The fundamental concept entails the development of contracts utilizing *transfer payments* between agents as a means to motivate them to engage in actions that optimize the total profits of the supply chain.

According to Cachon (2003), coordination of the supply chain occurs when the activities that lead to the best outcomes for the supply chain constitute Nash equilibrium, meaning that no individual business can gain profit by unilaterally deviating from the actions that are optimal for the supply chain. *Transfer payments* play a crucial role in facilitating coordination, and contracts can be categorized into various kinds based on their association with transfer payments. This section is motivated by Cachon (2003) and provides an overview of three prevalent supply chain contracts.

Consider a supply chain with two agents: a retailer (r) and a supplier (s). The retailer is a typical newsvendor who faces a stochastic demand $D > 0$ with Φ and ϕ as the cumulative distribution and probability density functions, respectively. Also, assume Φ is differentiable, strictly increasing, and $\Phi(0) = 0$. The retailer bears the inventory risk of the unsold units. Table 4.2 summarizes other parameters.

---

[9] This is succinctly captured in the following quote from the book "The Machine That Changed the World" (Womack et al., 2007)
Relations between the factory and the dealer were distant and usually strained as the factory tried to force cars on dealers to smooth production. Relations between the dealer and the customer were equally strained because dealers continually adjusted prices—made deals—to adjust demand with supply while maximizing profits.

Table 4.2 Model parameters

Notation	Description
c_s	Supplier's production cost per unit
c_r	Retailer's marginal cost per unit; incurred while procuring
g_s	Goodwill cost or backorder cost for the supplier
g_r	Goodwill cost or backorder cost for the retailer
$v < c_s + c_r$	Salvage value at the retailer level
$p > c_s + c_r$	Retail price

If retailer decides to order q units, then the expected sales $S(q) = \mathbb{E}[\min\{q, D\}] = q(1 - \Phi(q)) + \int_0^q x\phi(x)\,dy = q - \int_0^q \Phi(y)\,dy$. Similarly, the leftover inventory as a function of q is $I(q) = \mathbb{E}[q - D]^+ = q - S(q)$ and the lost-sales is $L(q) = \mathbb{E}[D - q]^+ = \mu - S(q)$, where $\mu = \mathbb{E}(D)$ denotes the mean demand. To formulate the profit functions of both retailer and supplier, let us assume that the supplier makes the contract offer—this assumption is made only for expositional convenience, and reversed roles do not affect the analysis.

For a generic contract, let T denote the transfer payment from retailer to supplier. The retailer's profit function is

$$\pi_r(q) = pS(q) + vI(q) - g_r L(q) - c_r q - T = (p - v + g_r)S(q) - (c_r - v)q - g_r \mu - T \quad (4.19)$$

Similarly, the supplier's profit function is

$$\pi_s(q) = g_s S(q) - c_s q - g_s \mu + T \quad (4.20)$$

and the total supply chain profit is

$$\pi(q) = \pi_s(q) + \pi_r(q) = (p - v + g_r + g_s)S(q) - (c_r + c_s - v)q - g\mu \quad (4.21)$$

If the supply chain is vertically integrated,[10] then the optimal order quantity q^* satisfies Eq. 4.22.

$$S'(q^*) = 1 - \Phi(q^*) = \frac{c_s + c_r - v}{p - v + g_s + g_r} \quad (4.22)$$

This expression can also be computed using the *critical fractile formula* from Eq. 1.3. Since Φ is strictly increasing, it means $\pi(q)$ is strictly concave, and q^* is unique.

The goal of supply contracts for coordination is to achieve the optimal total supply chain profit by using the optimal order quantity q^*. As mentioned earlier, this can be achieved by using transfer payments, which align the goals of the supply chain entities with the goal of maximizing the total supply profit.

---

[10] Assume it is controlled by the supplier who is now the newsvendor. This setting is not an exception. EV automaker Tesla directly sells to the customer (D2C model).

4.4.1 Wholesale-Price Contract

Wholesale-price contracts are agreements that determine the terms, pricing, and conditions for the bulk sale of products between manufacturers or suppliers and wholesale distributors or buyers (Lariviere & Porteus, 2001). These contracts facilitate the efficient exchange of goods in large quantities. They help establish pricing structures, minimum order quantities, delivery schedules, and other terms to benefit both the supplier and the buyer. These contracts ensure the availability of products and foster strong business relationships. Two real-life applications are provided in Fig. 4.10, but the list extends to various industries, such as manufacturing, construction, textiles, automotive, technology, and agriculture.

In these contracts, the supplier charges the retailer a fixed wholesale price of w per unit, and the transfer function $T(q, w)$ is wq. The retailer's profit function modifies to

$$\pi_r(q) = (p - v + g_r)S(q) - (c_r - v)q - g_r\mu - wq \tag{4.23}$$

Since Φ is strictly increasing, it means $\pi_r(q)$ is strictly concave, and the retailer's optimal order quantity q_r^* is unique and satisfies Eq. 4.24.

$$S'(q_r^*) = 1 - \Phi(q_r^*) = \frac{w + c_r - v}{p - v + g_r} \tag{4.24}$$

To keep the analysis simple, we assume $g_s = 0$. Using Eqs. 4.22 and 4.24, $q^* = q_r^*$ if and only if $w = c_s$. In other words, the retailer's optimal order quantity q_r^* matches the supply chain's optimal order quantity q^* if and only if the supplier's profit is zero ($w = c_s$ implies $\pi_s = 0$). It is rational for the supplier not to enter into a wholesale-price contract as coordination is achieved only with zero profit for the supplier.

We now discuss the contract from a game-theoretic viewpoint. Given w, q_r^* is the best response of the retailer and can be computed using Eq. 4.24. Knowing the best response function of the retailer, the supplier can compute the optimal wholesale price by substituting the best response function of the retailer in his profit function $\pi_s(q) = (w(q) - c_s)q$. It is assumed that $g_s = 0$. The supplier is willing to contract if the optimal wholesale price $w^* > c_s$. We already know that if $w^* > c_s$, then $q_r^* <$

Fig. 4.10 Applications of wholesale-price contracts

q^*. (q_r^*, w^*) is Nash equilibrium as unilateral deviations are suboptimal, and the wholesale-price contract is not coordinating the supply chain at Nash equilibrium.

Though wholesale-price contracts do not coordinate supply chains,[11] they are commonly used in practice due to their ease of implementation.

4.4.2 Buyback Contract

In the wholesale-price contract, the incentives are not *right* to coordinate the supply chain. Buyback contract improves the incentives and coordinates the supply chain. A *buyback contract* is a type of agreement where a retailer purchases products from a supplier and later has the option to sell any unsold inventory back to the supplier at a prearranged price or under specified conditions. This arrangement helps the retailer manage inventory risk and can be particularly useful for seasonal or perishable goods. If the retailer cannot sell all the products, they can return them to the supplier, reducing the risk of overstocking and potential financial losses. Buyback contracts provide a form of inventory management and risk-sharing between the supplier and the retailer.

Figure 4.11 shows two applications of buyback contracts: aircraft leasing and book publishing companies. In the aircraft leasing industry, a buyback contract may involve an airline leasing an aircraft from a manufacturer with a prearranged agreement that the manufacturer will repurchase the aircraft after a certain period or upon specified conditions. This allows the airline to use the aircraft temporarily without committing to full ownership. In the book publishing industry, a buyback contract may refer to agreements between publishers and bookstores. Publishers agree to buy back unsold books from bookstores, typically at a reduced price, to help bookstores manage inventory and reduce financial risk in the event of unsold copies.

In addition to the wholesale-price contract with the wholesale price w_b, the supplier compensates b per unit for unsold items at the retailer level. The terminology employed may potentially lead to confusion, as it conveys the notion that the remaining items are reclaimed by the provider. This situation may arise when the salvage value of the supplier exceeds that of the retailer. Furthermore, it is important to acknowledge that the expense associated with returning the remaining items to the supplier significantly impacts the analysis, thereby adding complexity without providing substantial valuable findings. Therefore, it is postulated that the salvage value of the retailer is greater, and the supplier possesses the technological capabilities to accurately monitor the quantity of unsold goods remaining in the retailer's establishment. Given this setting, the corresponding transfer payment function is

$$T = T(q, w_b, b) = w_b q - b I(q) = b S(q) + (w_b - b)q \qquad (4.25)$$

---

[11] Vipin and Amit (2021) show that a wholesale-price contract can coordinate a supply chain with a behavioral retailer under certain conditions.

4.4 Supply Chain Contracts

Fig. 4.11 Applications of buyback contracts

which modifies the retailer's profit function to

$$\pi_r(q, w_b, b) = (p - v + g_r - b)S(q) - (w_b - b + c_r - v)q - g_r\mu \quad (4.26)$$

Since Φ is strictly increasing, it means $\pi_r(q, w_b, b)$ is strictly concave, and the retailer's optimal order quantity $q^*_{r,bb}$ is unique and satisfies Eq. 4.27.

$$S'(q^*_{r,bb}) = \frac{w_b - b + c_r - v}{p - v + g_r - b} \quad (4.27)$$

The supply chain is coordinated if $q^*_{r,bb}$ from Eq. 4.27 is equal to q^* from Eq. 4.22. To keep the analysis simple, we assume $g_s = g_r = 0$. Assume $p - v - b = \lambda(p - v)$ and $w_b - b + c_r - v = \lambda(c_r + c_s - v)$. This modification ensures $q^*_{r,bb} = q^*$, and the retailer's profit function changes to

$$\pi_r(q, w_b, b) = \lambda \pi(q) \quad (4.28)$$

The resulting supplier's profit function is

$$\pi_s(q, w_b, b) = \pi(q) - \pi_r(q, w_b, b) = (1 - \lambda)\pi(q) \quad (4.29)$$

From Eqs. 4.28 and 4.29, it is evident that parameter λ divides the supply chain profit between the retailer and the supplier. As long as $\lambda \leq 1$, the buyback contract coordinates the supply chain. However, the multiplicity of λ means that there are multiple Nash equilibria that coordinate the supply chain. Devangan et al. (2013) discuss individually rational buyback contracts to reduce the set of Nash equilibria.

4.4.3 Revenue-Sharing Contract

In the wholesale-price contract, we showed that the supply chain coordinates if $w = c_s$. However, with $w = c_s$, the supplier's profit is zero, and the supplier may not participate in the contract. One option to ensure the participation of the supplier, with

Fig. 4.12 Applications of revenue-sharing contracts

$w = c_s$, is that the retailer shares the revenue generated from the sales of products with the supplier. Such a class of contracts is called a *revenue-sharing contract*. Revenue-sharing contracts between a supplier and retailer are agreements where both parties share a portion of the revenue generated from the sale of products (Cachon & Lariviere, 2005). This arrangement allows both the supplier and the retailer to have a vested interest in the success of the products, as their compensation is tied to the actual sales or revenue generated, fostering a collaborative and mutually beneficial relationship.

Streaming platforms like Netflix and Amazon Prime and rental outlets in malls serve as two real-world applications of revenue-sharing contracts, as shown in Fig. 4.12. In the context of Netflix and Amazon Prime, revenue-sharing contracts typically involve content creators, such as production companies or filmmakers, sharing a portion of the subscription or viewing fees with the streaming platforms. This arrangement is based on the popularity and viewership of their content. In rental outlets in malls, revenue-sharing contracts often occur between mall owners and individual rental businesses. Mall owners receive a percentage of the rental business's sales revenue, allowing both parties to benefit from the mall's foot traffic and overall success.

Assuming that the retailer's revenue share includes the portion of salvage value, the transfer payment for this contract is given by

$$T = T(q, w_r, \psi) = w_r q + (1 - \psi) v I(q) + (1 - \psi) p S(q)$$
$$= (w_r + (1 - \psi)v)q + (1 - \psi)(p - v)S(q)$$

where ψ is the portion of the revenue the retailer keeps with himself. w_r is the wholesale price per unit in the contract. With this, the retailer's profit function changes to

$$\pi_r(q, w_r, \psi) = (\psi(p - v) + g_r)S(q) - (w_r + c_r - \psi v)q - g_r \mu \qquad (4.30)$$

Since Φ is strictly increasing, it means $\pi_r(q, w_r, \psi)$ is strictly concave, and the retailer's optimal order quantity $q^*_{r,rs}$ is unique and satisfies Eq. 4.27.

4.5 Blockchains

$$S'(q_{r,rs}^*) = \frac{w_r + c_r - \psi v}{\psi(p - v) + g_r} \tag{4.31}$$

The supply chain is coordinated if $q_{r,rs}^*$ from Eq. 4.31 is equal to q^* from Eq. 4.22. To keep the analysis simple, we assume $g_s = g_r = 0$. Assume $\psi(p - v) = \lambda(p - v)$ and $w_r + c_r - \psi v = \lambda(c_r + c_s - v)$. This modification ensures $q_{r,rs}^* = q^*$, and the retailer's profit function changes to

$$\pi_r(q, w_r, \psi) = \lambda \pi(q) \tag{4.32}$$

The resulting supplier's profit function is

$$\pi_s(q, w_r, \psi) = \pi(q) - \pi_r(q, w_r, \psi) = (1 - \lambda)\pi(q) \tag{4.33}$$

Equations 4.32 and 4.33 are similar to Eqs. 4.28 and 4.29. This establishes a connection between buyback and revenue-sharing contracts. Similar to buyback contracts, when $\lambda \leq 1$, the revenue-sharing contract coordinates the supply chain. However, the multiplicity of λ means that there are multiple Nash equilibria that coordinate the supply chain.

In this section, we discussed various types of contracts for supply chain coordination. Though wholesale-price contracts do not coordinate supply chains, they are easy to implement. In other contracts, additional effort is required from the supplier in order to observe and verify the information pertaining to inventory surplus, sales figures, and market demand. Using the nomenclature from mechanism design from Chap. 9, wholesale-price contracts are *detail-free*, and hence easy to implement.

Contract theory is a vast research area with numerous applications. The literature on supply chain coordination is limited to a small subset of contract theory, and much research is needed to connect supply chain contracts to the other dimensions of contract theory. Bolton and Dewatripont (2004) is an authoritative and detailed source on contract theory.

4.5 Blockchains[12]

During the initial stages of the computer revolution, the high cost of computers necessitated storing and managing data at a single location, referred to as a *centralized database*. These databases provided guarantees of data integrity and portability and also offered simplified administration processes. One example is financial institutions, such as banks, which retain all transactional data on their servers. Nevertheless, these databases were susceptible to attacks and presented an elevated vulnerability to data manipulation and theft due to their centralized management by a sole governing entity at a single physical location.

---

[12] In this section, the OM actions are on the *process side* in the VCAP framework.

As the cost of computers has decreased over time, advancements in technology have facilitated the development of *distributed databases*, enabling the storage of data across several physical locations.[13] The network's nodes can generate, access, and alter data from any computer linked to the database. Moreover, because of the distribution of data across the network, these nodes exhibit a reduced susceptibility to attacks. Nevertheless, the entirety of the set is predicated on the underlying premise that all nodes consistently adhere to honesty, refraining from any manipulation of data for personal gain. This assumption holds validity to a certain degree, but its applicability is limited in cases when the nodes exhibit self-interest, rationality, and strategic behavior. Furthermore, some nodes may be malicious and can engage in *non-credible cheap talk* (see Sect. 3.4.4.1). These malicious nodes can impact the trust in the information in the network. Is it possible to design a mechanism to achieve a consensus about the true state of the information in a network when nobody can be completely sure who can be trusted?[14] This has long been recognized as a problem in the field of computer science—*The Byzantine Generals Problem*, and such mechanisms are called *trustless consensus-building mechanisms*.

Designing such mechanisms needs multiple technological innovations—*blockchains, digital signatures*, and *cryptographic hash puzzles*. Cryptocurrencies like Bitcoin use these technological ideas to design trustless consensus mechanisms to make any changes in the distributed ledger. We recommend the original paper on Bitcoins by Nakamoto (2008) as the best resource to understand the Bitcoin protocol. Our further discussion on blockchains is motivated by the Bitcoin protocol. Despite blockchain implementation's inherent instability and nascent nature, Bitcoin, as one of its applications, has demonstrated sustained functionality for over a decade (Babich & Hilary, 2020).

Distributed databases in Bitcoin architecture are called *decentralized ledgers*. In Bitcoin architecture, a blockchain, introduced by Haber and Stornetta (1991), is a decentralized ledger consisting of discrete blocks that are linked sequentially and overseen by a peer-to-peer network. A *cryptographic hash pointer*[15] can be employed to combine the records included within the said block. The hash pointer of a block is connected to the subsequent block, enabling the hash pointer to encompass all previous entries within the chain. This relationship is visually depicted in Fig. 4.13.

---

[13] Three technology laws, *Moore's law*, *Gilder's law*, and *Metcalfe's law* are the catalyst in this transition.

[14] This has a flavor of mechanism design discussed in Chap. 9.

[15] A cryptographic hash pointer is a *cryptographic hash function* that produces a fixed size output for input of any size and is efficiently computable. Furthermore, it has properties of *collision resistance*, *hiding*, and *puzzle-friendliness*.

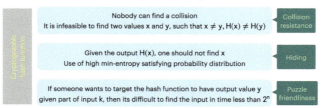

4.5 Blockchains

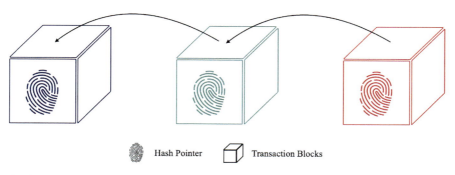

Fig. 4.13 Blockchain

Fig. 4.14 Properties of a blockchain

In this decentralized system, copies of the ledger are available to each participant, and a process is established by which participants agree to make any changes in the ledger (that is, on which transactions are valid). Due to their numerous benefits, blockchains are employed in several domains, such as supply chains, capital markets, cryptocurrencies, and the Internet of Things (IoT). Furthermore, prospective future uses encompass election voting and tax monitoring within governmental contexts, healthcare, and energy sectors in business, as well as cybersecurity. Some of the properties of blockchains are shown in Fig. 4.14.

Bitcoin operates on a public blockchain, wherein all nodes maintain anonymity. *Digital signatures* provide mathematical proof that a particular transaction was done by a particular person. Digital signatures are an application of public-key cryptography, which relies on two separate keys: private and public. The keys are mathematically related. Bitcoin grants each node a pair of keys, namely, private and public keys. The private key is only known to the node and is maintained confidentially, while the public key can be utilized during the processing of transactions. The design of these systems allows for the generation of a significant quantity of distinct public keys through the utilization of the private key. However, the reverse process is not feasible. The utilization of the public key as the node's identity ensures anonymity.

As mentioned earlier, Bitcoin uses a consensus mechanism by which users agree on changes to the ledger. The consensus mechanism should ensure that the changes

80   4 Games in Normal Form: Applications in OM

must be proposed by an honest node rather than a malicious node. This is akin to removing non-credible cheap talk that can impact the trust in the information in the network. The consensus mechanism uses the *Proof-of-Work* method, which involves solving *cryptographic hash puzzles*. Dwork and Naor (1992) propose the concept of hash puzzles to mitigate the issue of email spam. The process of solving cryptographic problems is commonly referred to as *Bitcoin mining*, with the nodes in this process being referred to as *miners*. A typical hash puzzle is

$$\text{Given } \mathsf{H(nonce||prevhash||tx||tx||...||tx)} < \mathsf{target}, \quad \text{find nonce value}$$

Due to the puzzle-friendliness property of hash functions, the nonce value cannot be computed but can be guessed, which needs substantial computing power. Furthermore, the complexity of solving the hash puzzle is increasing with growing computational capabilities (Nakamoto, 2008). The proof-of-work approach involves the node that successfully solves the cryptographic puzzle being designated to propose the subsequent block in the blockchain. The proposal undergoes a process of validation by peers, and upon receiving their approval, the block is appended to the chain. In Bitcoin, the puzzles are designed to add a block approximately every 10 minutes. Bitcoin protocol requires that the miners contributing to the verification process must demonstrate a cryptographic proof of work to show that they have paid a cost in computation time before their proposals are accepted, hence making the cheap talk *expensive*. Blockchains lower transaction costs[16] through costless verification and without the need for costly intermediation. For further reading, Narayanan et al. (2016) is a good source on Bitcoin and blockchains.

With this brief introduction to blockchains, we dive into the game-theoretic dimensions relevant to blockchains. We discussed its relevance in mitigating non-credible cheap talk. The other connection is *mining pools* for collaborative mining. Mining pools have been used since 2010, yet they were unable to remain cooperative because of the incentive differences. Mining pools are one area where the practice lies ahead of theory, and cooperative game theory (discussed in Chap. 7) can be used to study mining pools. Lewenberg et al. (2015) is one such study.

Bitcoin mining is a simultaneous game where miners spend expensive computing resources. It is important to model the strategic behavior of the miners to improve consensus-building mechanisms. Dimitri (2017) models the Bitcoin mining activity as *bitcoin mining game*. As the mining activity is performed by all the nodes, but only one wins, the author models it as an all-pay contest with a probabilistic victory. Suppose there are n active miners in the network, and a miner i chooses to invest $h_i > 0$ in getting the computing power. The winner gets a reward $R \geq 0$ for successful mining. For the time being, consider R as the block reward in terms of bitcoins. Let X_i be an exponentially distributed random variable, which denotes the miner i's waiting time to solve the puzzle with parameter h_i/d, where d is the difficulty level indicator and gets adjusted based on the network's total computing power. Assuming that returns to scale are constant with the investment, the cost incurred for mining

---

[16] Recall technologies that impact transaction costs in Fig. 4.9.

4.5 Blockchains

is $c_i h_i$, where c_i is the marginal cost for acquiring one unit of computational power. With this, the miner i profit as a function of computing power $\pi_i(h_i)$ is given by

$$
\pi_i(h_i) = \begin{cases} R - c_i h_i & & \frac{h_i}{h_{(n)}} & h_i > 0 \\ -c_i h_i & \text{with probability} & \frac{h_{-i}}{h_{(n)}} & \text{if } h_i > 0 \\ 0 & & 1 & h_i = 0 \end{cases}
$$

where $h_{(n)} = \sum_{i=1}^{n} h_i$ and $\frac{h_i}{h_{(n)}}$ is the probability that miner i is the winner. Then, the miner's expected profit is given by

$$
\mathbb{E}[\pi_i(h_i)] = \frac{R h_i}{h_{(n)}} - c_i h_i, \quad i = 1, 2, .., n. \tag{4.34}
$$

Bitcoin mining game is normal-form game with the tuple, $\langle N = \{i : i = \{1, 2\}\}, a_i = h_i \in [0, \infty), u_i = \mathbb{E}[\pi_i(h_i)]$.

First-order conditions of Eq. 4.34 with respect to h_i yield

$$
\frac{R h_{-i}}{h_{(n)}^2} = c_i \tag{4.35}
$$

where $h_{-i} = h_{(n)} - h_i$. From Eq. 4.35, it is obvious that $\partial^2 \mathbb{E}[\pi_i(h_i)]/\partial h_i^2 < 0$, i.e., the expected profit is strictly concave with respect to h_i. Furthermore, using Eq. 4.35, it can be shown that

$$
h_{(n)} = \frac{R(n - 1)}{c_{(n)}} \tag{4.36}
$$

where $c_n = \sum_{i=1}^{n} c_i$.

In Eq. 4.35, the optimal h_i^* for miner i can be computed as

$$
h_i^* = h_{(n)} - \frac{h_{(n)}^2 c_i}{R} = \frac{h_{(n)}[c_{(n)} - (n - 1)c_i]}{c_{(n)}} = \frac{R(n - 1)[c_{(n)} - (n - 1)c_i]}{c_{(n)}^2} \tag{4.37}
$$

Without loss of generality, we assume $c_1 \leq c_2 \leq c_3 \leq ... \leq c_n$ which implies $h_1^* \geq h_2^* \geq ... \geq h_n^*$. The above implications summarize the main result in Proposition 4.2.

Proposition 4.2 (Dimitri, 2017) *The unique pure strategy Nash equilibrium of the Bitcoin mining game is the profile $(h_1^* \geq h_2^* \geq ... \geq h_n^*)$, where h_i^* is as per Eq. 4.37.* ◀

Proposition 4.2 has two interesting observations. Firstly, the decision to be an active miner ($h_i^* > 0$) is dependent only on the miner's cost structure when compared with other miners because the necessary condition is that $c_{(n)} - (n - 1)c_i > 0$. Note that the mining activity is independent of the reward R as long as it is strictly positive. Secondly, substituting Eqs. 4.35 and 4.37 in Eq. 4.34 yields

$$\mathbb{E}[\pi_i(h_i)] = \frac{Rh_i^2}{h_{(n)}^2} \tag{4.38}$$

which means that the expected profits are a function of the block reward R.

Also, the expected rate of returns $r_i(h_i)$ can be derived as

$$r_i(h_i) = \frac{\mathbb{E}[\pi_i(h_i)]}{c_i h_i} = \frac{c_{(n)} - (n-1)c_i}{(n-1)c_i} > 0 \tag{4.39}$$

$$> \frac{c_{(n)} - (n-1)c_i}{c_{(n)}} = \frac{h_i}{h_{(n)}} \tag{4.40}$$

This suggests that the expected rate of returns exceeds the likelihood of winning and is not influenced by the block reward. These two facts elucidate the reasons behind the continued existence of Bitcoin despite the diminishing block reward. Furthermore, it is noteworthy to observe that miners' involvement primarily relies on their opponents' computational capabilities rather than the incentives received for achieving successful mining outcomes.

Problems

Problem 4.1 A wholesale contract is negotiated between a supplier and a retailer for a certain product. The retailer plans to sell this product to the end customers at a retail price of ₹100. The variable cost associated with producing this product is ₹30, and the supplier's gross margin is 20% of the retail price, while the retailer's gross margin is 30% of the retail price. The supplier has a fixed cost of ₹1,000, and the retailer has a fixed cost of ₹2,000.

   i Discuss the economic implications of the calculated wholesale price for both the supplier and the retailer. What factors influence the wholesale price, and how do they affect the profitability of each party in the contract?

   ii Imagine the retailer negotiates with the supplier to reduce the retailer's fixed cost by ₹500. How does this change impact the wholesale price, and what are the implications for the profitability of both the supplier and the retailer? Provide a new wholesale-price calculation and a brief analysis. ◀

Problem 4.2 Suggest the most appropriate supply chain contract with justification for each supply chain setting.

   i An industry with long production cycles where stability is essential.

   ii A situation with high demand uncertainty and a need to minimize carrying costs.

iii A logistics and transportation company aiming to ensure on-time deliveries.

iv A scenario involving a collaborative project with shared risks and rewards.

v A buyer who wants to secure a dedicated supplier for critical components.

vi An industry with frequent changes in demand and a focus on cost efficiency.

vii A situation with significant external risks and uncertainties.

viii A construction project with unpredictable costs and a focus on quality.

ix An agricultural sector where unsold produce needs to be managed efficiently.

x A distribution center handling goods for retail stores with minimal storage space. ◀

Problem 4.3 For every 210,000 blocks mined, which is roughly every 4 years, the reward that miners receive is halved. When Bitcoin was first created, the block reward was 50 bitcoins. After the first halving in 2012, it was reduced to 25 bitcoins. Incorporate this dependency in the Bitcoin reward function and compute Nash equilibrium for the mining game (refer Sect. 4.5). Interpret the results. ◀

References

Babich, V., & Hilary, G. (2020). Om forum—distributed ledgers and operations: What operations management researchers should know about blockchain technology. *Manufacturing & Service Operations Management, 22*(2), 223–240.

Belobaba, P. P. (1989). Or practice—application of a probabilistic decision model to airline seat inventory control. *Operations Research, 37*(2), 183–197.

Bolton, P., & Dewatripont, M. (2004). *Contract theory.* MIT Press.

Cachon, G. P. (2003). *Supply chain coordination with contracts.* Handbooks in operations research and management science (Vol. 11, pp. 227–339). New York: Springer.

Cachon, G. P., & Lariviere, M. A. (2005). Supply chain coordination with revenue-sharing contracts: Strengths and limitations. *Management Science, 51*(1), 30–44.

Coase, R. H. (1937). The nature of the firm. *Economica, 4,* 386–405.

Devangan, L., Amit, R. K., Mehta, P., Swami, S., & Shanker, K. (2013). Individually rational buyback contracts with inventory level dependent demand. *International Journal of Production Economics, 142*(2), 381–387.

Dimitri, N. (2017). Bitcoin mining as a contest. *Ledger, 2,* 31–37.

Dwork, C., & Naor, M. (1992). Pricing via processing or combatting junk mail. In *Annual International Cryptology Conference* (pp. 139–147). Springer.

Haber, S., & Stornetta, W. S. (1991). *How to time-stamp a digital document.* Springer.

Hu, X., Caldentey, R., & Vulcano, G. (2013). Revenue sharing in airline alliances. *Management Science, 59*(5), 1177–1195.

Joe, H. (1997). *Multivariate models and multivariate dependence concepts.* CRC Press.

Knödel, W. (2013). *Graphentheoretische methoden und ihre anwendungen* (Vol. 13). Springer.

Kolata, G. (1990). What if they closed 42d street and nobody noticed. *New York Times, 25,* 38.

Kreps, D. M. (2004). *Microeconomics for managers.* New York: Norton.

Lariviere, M. A., & Porteus, E. L. (2001). Selling to the newsvendor: An analysis of price-only contracts. *Manufacturing & Service Operations Management, 3*(4), 293–305.

Lewenberg, Y., Bachrach, Y., Sompolinsky, Y., Zohar, A., & Rosenschein, J. S. (2015). Bitcoin mining pools: A cooperative game theoretic analysis. In *Proceedings of the 2015 International Conference on Autonomous Agents and Multiagent Systems* (pp. 919–927).

Littlewood, K. (1972). Forecasting and control of passenger bookings. In *Airline Group International Federation of Operational Research Societies Proceedings, 1972* (Vol. 12, pp. 95–117).

Mcgillivray, R., & Silver, E. (1978). Some concepts for inventory control under substitutable demand. *INFOR: Information Systems and Operational Research, 16*(1), 47–63.

Murchland, J. D. (1970). Braess's paradox of traffic flow. *Transportation Research, 4*(4), 391–394.

Nagurney, A., & Nagurney, L. S. (2021). The Braess paradox. *International encyclopedia of transportation*. Elsevier.

Nakamoto, S. (2008). Bitcoin: A peer-to-peer electronic cash system. *Decentralized Business Review*, 21260.

Narayanan, A., Bonneau, J., Felten, E., Miller, A., & Goldfeder, S. (2016). *Bitcoin and cryptocurrency technologies: A comprehensive introduction.* Princeton University Press.

Netessine, S., & Shumsky, R. A. (2005). Revenue management games: Horizontal and vertical competition. *Management Science, 51*(5), 813–831.

Parlar, M. (1988). Game theoretic analysis of the substitutable product inventory problem with random demands. *Naval Research Logistics (NRL), 35*(3), 397–409.

Parlar, M., & Goyal, S. (1984). Optimal ordering decisions for two substitutable products with stochastic demands. *Opsearch, 21*(1), 1–15.

Pas, E. I., & Principio, S. L. (1997). Braess' paradox: Some new insights. *Transportation Research Part B: Methodological, 31*(3), 265–276.

Pearson, J. (2022). 25 years on: Inside the three global airline alliances.

Schulz, A. S., & Stier Moses, N. E. (2002). On the performance of user equilibrium in traffic networks.

Spengler, J. J. (1950). Vertical integration and antitrust policy. *Journal of Political Economy, 58*(4), 347–352.

Van Mieghem, J. A., & Allon, G. (2015). *Operations strategy*. Belmont, MA: Dynamic Ideas.

Vipin, B., & Amit, R. K. (2021). Wholesale price versus buyback: A comparison of contracts in a supply chain with a behavioral retailer. *Computers & Industrial Engineering, 162*, 107689.

Williamson, O. E. (1995). *The mechanisms of governance*. New York: Oxford University Press.

Womack, J. P., Jones, D. T., & Roos, D. T. A. T. T. (2007). The machine that changed the world.

Wu, H., & Parlar, M. (2011). Games with incomplete information: A simplified exposition with inventory management applications. *International Journal of Production Economics, 133*(2), 562–577.

Chapter 5
Games in Extensive Form

Chapter 3 detailed the noncooperative games in which the players move simultaneously. Another way in which many strategic actions arise in everyday life, as well as in economics, involves players moving sequentially. For example, an investment made today may produce certain strategic advantages in the future. Such situations are modeled using *extensive-form games*. We discussed extensive-form games in Sect. 2.2.2. We start with some examples to demonstrate the elegance with which temporal problems can be represented with the help of an extensive-form game tree.

5.1 Examples

Example 5.1 (*Entry Deterrence Game*) Consider a situation in which a new firm contemplates entering an existing market. Naturally, the entrant firm ($\mathscr{E}$) becomes the player to make the first move in the game. In case $\mathscr{E}$ decides to choose Enter (E), the incumbent ($\mathscr{I}$) firm faces a dilemma. It can Fight (F) by aggressively pricing the product or Acquiesce (A) by accepting duopoly pricing. Alternatively, $\mathscr{E}$ can also choose to stay out of the market by choosing Don't Enter (DE). Figure 5.1 represents the situation as extensive-form game.

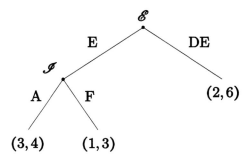

Fig. 5.1 Entry deterrence game

It is obvious that $\mathscr{I}$ prefers $\mathscr{E}$ to stay out of the market. While if $\mathscr{E}$ decides to enter the market, $\mathscr{I}$ prefers to choose A. It is convenient at this point to invoke the notion of rationality discussed in Chap. 2. Firm $\mathscr{E}$ is intelligent enough to connect the dots, and conjectures that the best response of $\mathscr{I}$, conditional on the choice of E by $\mathscr{E}$, is A. Hence, a reasonable outcome of the game is (E, A) with payoff (3, 4).

Now suppose, the incumbent firm $\mathscr{I}$ indulges in a costless marketing campaign (recall preplay communication in Sect. 3.4.4) to keep $\mathscr{E}$ out of the market by declaring to fight for the market share it holds. Should $\mathscr{E}$ believe in this communication? Is the threat of $\mathscr{I}$ to choose F once $\mathscr{E}$ enters the market credible? The answer is *no* because $\mathscr{E}$ can reason out that $\mathscr{I}$ gets a higher payoff by choosing A than F once $\mathscr{E}$ decides to enter. The only reasonable outcome of the game is, in fact, (E, A). ◁

Example 5.2 (*Modified Entry Deterrence Game*) Consider a modification of the entry deterrence game. Firm $\mathscr{E}$ prepares for the scenario if $\mathscr{I}$ chooses F after the entry of $\mathscr{E}$. In this case, $\mathscr{E}$ can either Retaliate (R) or Leave (L) the market. Extensive-form representation of the situation is shown in Fig. 5.2. What are the strategies of $\mathscr{E}$ and $\mathscr{I}$? We discuss strategies for extensive-form games in Sect. 5.2.

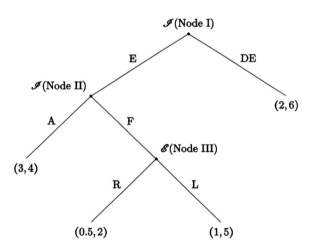

Fig. 5.2 Modified entry deterrence game ◁

Example 5.3 (*von Stackelberg Duopoly*) We discussed Cournot duopoly in Example 3.3. In the Cournot duopoly game, each firm has *Cournot conjecture* that the other firm will compete by choosing a fixed quantity of a substitutable good. von Stackelberg (1934) modifies the setting in which Firm 1 makes the first move and chooses the quantity a_1. Firm 1 believes that Firm 2 has Cournot conjecture. It means that Firm 2 decides its optimal quantity based on the choice of a_1. The setting is called the *von Stackelberg duopoly*. Firm 1 is the *von Stackelberg leader*, and Firm 2 is the *von Stackelberg follower*.

5.1 Examples

Using similar notations as in Example 3.3, extensive-form representation of the setting is shown in Fig. 5.3. Firm i maximizes its profit function $u_i = (10 - a_1 - a_2) \times a_i - c \times a_i$ by choosing a production quantity a_i, with the market price as $(10 - a_1 - a_2)$.

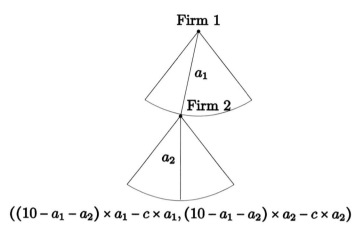

Fig. 5.3 Stackelberg game

Observing the production quantity a_1 of Firm 1, Firm 2 selects its production quantity a_2 that maximizes its utility. Maximizing u_2 for a given a_1 results in $a_2^*(a_1) = \frac{10-c-a_1}{2}$. This is the best response function of Firm 2. Firm 1 can conjecture this response from Firm 2 and uses this response function as an input to solve its utility maximization problem given by

$$\max_{a_1} \; (10 - a_1 - a_2) \times a_1 - c \times a_1$$

$$\max_{a_1} \; \left(10 - a_1 - \left(\frac{10 - c - a_1}{2}\right)\right) \times a_1 - c \times a_1$$

Solving these equations, the equilibrium quantities are $a_1^* = \frac{10-c}{2}$ and $a_2^* = \frac{10-c}{4}$. The von Stackelberg leader has the first-mover advantage as it captures a greater market share than the follower firm. Compare the outcomes of the von Stackelberg duopoly with that of the Cournot duopoly where the equilibrium quantity is $a_i^* = \frac{10-c}{3}$ for each firm. ◁

Example 5.4 (*Finite Horizon Bargaining Game*) Game theory can be used to resolve conflicts in which two parties would rather settle on some sort of agreement, giving them a greater payoff than complete disagreement. A pertinent example is a typical bargaining between a buyer and a seller in a marketplace. Suppose two players bargain to distribute the value of trade ₹100 between themselves. They use *alternating offer bargaining protocol*. In this protocol, in round 1, Player 1 offers to keep some amount

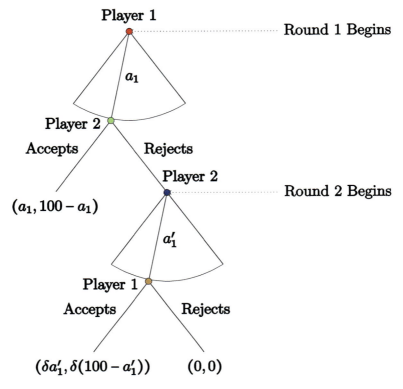

Fig. 5.4 Finite-horizon bargaining game

$a_1 \in [0, 100]$ for herself and $(100 - a_1)$ for Player 2. Player 2 can either *accept* the offer and the bargaining stops, or *reject* the offer, which leads to the next round of bargaining. In round 2, player 2 offers the division $(\delta a_1', \delta(100 - a_1'))$, where δ is the discount factor applied as a penalty for extending the game to the second round.[1] Player 1 can either accept Player 2's offer or reject the offer. There are no further rounds of bargaining. The extensive-form representation is shown in Fig. 5.4.

The game can be solved using *backward induction*, which is an iterative procedure of optimizing in the final choice node and then moving backward to determine the optimal sequence of actions. In the finite-horizon bargaining game, Player 1 makes the final move at the brown node, and Player 2 can conjecture that Player 1 will accept the offer at the brown node if $\delta a_1' \geq 0$, or $a_1' \geq 0$. Player 2 maximizes his payoff by offering $a_1' = 0$ to Player 1 at the blue node. Player 1, when offering the distribution at the red node, is rational enough to reason out that Player 2 at the green node will accept the offer only if $(100 - a_1) \geq \delta 100$, or $a_1 \leq 100(1 - \delta)$. Player 1's payoff is maximized by choosing $a_1 = 100(1 - \delta)$ for herself and offering $100 - a_1 = \delta 100$ to Player 2. Player 2, when moving at the green node, cannot do better by rejecting

---

[1] Discount factor captures the diminished value of money in future.

Player 1's offer. It should be noted that the solution depends on the discount factor δ. If there is no discounting of the payoffs, Player 2 has the advantage of making the last offer.

The finite-horizon version of the alternating offer bargaining protocol is attributed to Ståhl (1972). Rubinstein (1982) models the infinite-horizon version of the protocol.

◁

Example 5.5 (*Prisoners' Dilemma Revisited*) How can the Prisoners' Dilemma, discussed in Example 3.1 be represented as an extensive-form game? In the Prisoners' Dilemma, both the prisoners are unaware of the move made by the other prisoner. Alternatively, neither prisoner has any information regarding the move made by the other prisoner. Figure 5.5 represents the prisoners' dilemma setting as an extensive form game.

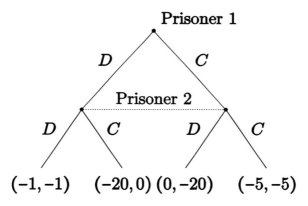

Fig. 5.5 Prisoners' dilemma as extensive-form game

We can assume either prisoner to be the first mover (Prisoner 1 here). Prisoner 2 is uncertain about the choice made by Prisoner 1. In other words, Prisoner 2 has imperfect information about the choices of Player 1, and hence, Prisoner 2's information set has two nodes. This is an extensive-form representation of the normal-form game in Example 3.1.[2]

◁

Example 5.6 (*Entry Deterrence Game with Incomplete Information*) Consider a setting in which the decision of Firm ($\mathscr{E}$) to the market with an incumbent Firm ($\mathscr{I}$) depends on the set-up cost incurred by $\mathscr{E}$. Assume that there are only two possible set-up costs—Low and High, each equally likely. Once the set-up costs are revealed to Firm $\mathscr{E}$, it decides whether to enter. If Firm $\mathscr{E}$ decides to enter, Firm $\mathscr{I}$ is not sure whether it is competing against Firm $\mathscr{E}$ with high set-up cost or low set-up cost. This is a setting with incomplete information and can be modeled as extensive-form game with imperfect information as in Fig. 5.6.

---

[2] For any extensive-form representation, there is a unique normal-form representation; however, the converse is not true (Mas-Colell et al., 1995).

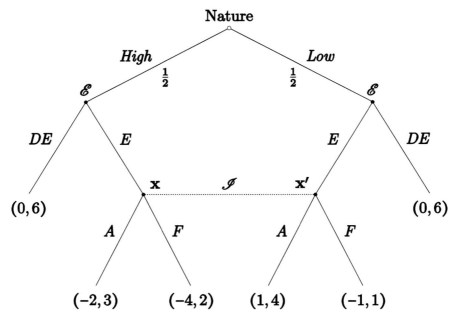

Fig. 5.6 Entry deterrence game with incomplete information

In this game, nature makes the first move and chooses Firm $\mathscr{E}$'s set-up cost. On observing nature's move, Firm $\mathscr{E}$ can decide whether to enter or not. But, Firm $\mathscr{I}$ does not observe the choice of nature. So if Firm $\mathscr{E}$ decides to enter, Firm $\mathscr{I}$ has imperfect information whether it is at node **x** or **x'**. ◁

Example 5.7 (*Rock–Paper–Scissors Game*) Consider a situation where two players P_1 and P_2 are engaged in Rock–Paper–Scissors game. P_1 unintentionally sends a signal that informs P_2 that P_1 is choosing Rock (R). P_2 cannot detect any signal when P_1 chooses Scissors (S) or Paper (P). Extensive-form representation of the setting is shown in Fig. 5.7. ◁

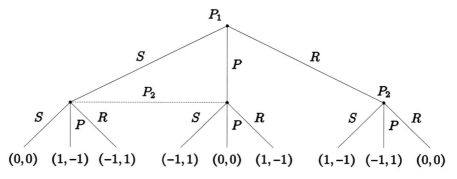

Fig. 5.7 Rock–Paper–Scissors game with imperfect information

5.2 Strategies in Extensive-Form Games

In this section, we formalize the concept of strategies for extensive-form games. *In extensive-form games, a strategy of a player is a mapping of each information set of the player to the available actions at the respective information set*, formally defined in Definition 5.1.

Definition 5.1 (*Pure Strategies in Extensive-form Games*) A strategy s_i for player i is defined as $s_i : I_i \mapsto A_i^K$, where I_i is the *set* of information sets of player i and $K \in I_i$ is an information set with A_i^K is the action set at K. ◄

In other words, a strategy in extensive-form games recommends actions at each information set, even if some of the information sets are not reached when the game is actually played. This way of defining strategies for extensive-form games is important for the concept of *sequential rationality*, which is essential for defining reasonable solution concepts for extensive-form games. Sequential rationality means that the players choose optimally at any information set, given the strategies of the other players from that information set onwards.[3] In Example 5.4, Player 2, when deciding at the green node, chooses "Accepts" because he cannot gain by rejecting the offer at the green node. His payoff by accepting the offer equals his sequentially rational payoff based on $a_1' = 0$ at the blue node. It is important to note that the blue node is not reached if Player 2 chooses "Accepts" at the green node.

The backward induction procedure can be used to compute sequentially rational strategies in finite extensive-form games[4] of perfect information. This is illustrated in Example 5.8.

Example 5.8 (*Backward Induction in Modified Entry Deterrence Game*) Let us revisit the Modified Entry Deterrence Game (Example 5.2).

What are the possible strategies for Firm $\mathscr{E}$ and Firm $\mathscr{I}$? Firm $\mathscr{I}$ has only one decision node: Node II (Fig. 5.8a), and there are two possible strategies for Firm $\mathscr{I}$:

$$s_{\mathscr{I}}^1 = \mathrm{A}; \quad s_{\mathscr{I}}^2 = \mathrm{F}$$

Firm $\mathscr{E}$ has two decision nodes: Node I and Node III. The four possible strategies for Firm $\mathscr{E}$ are

$$s_{\mathscr{E}}^1 = (\mathrm{E, R}); \quad s_{\mathscr{E}}^2 = (\mathrm{E, L}); \quad s_{\mathscr{E}}^3 = (\mathrm{DE, R}); \quad s_{\mathscr{E}}^4 = (\mathrm{DE, L})$$

Strategy $s_{\mathscr{E}}^4 = (\mathrm{DE,L})$ recommends Firm $\mathscr{E}$ to choose "DE" at Node I and "L" at Node III. With this strategy, Node III is not reached in the game's actual play. However, choosing "L" is sequentially rational at Node III for Firm $\mathscr{E}$. The sequentially rational choice for Firm $\mathscr{I}$ at Node II is "F", and hence the sequentially rational choice for Firm $\mathscr{E}$ at Node I is "DE". Sequentially rational choice at each node is shown as

---

[3] Sequential rationality captures folk wisdom—*today is the first day of rest of my life*.

[4] In finite extensive-form games, there are finite decision nodes.

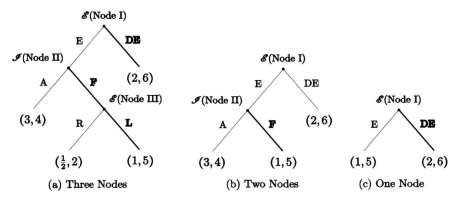

Fig. 5.8 Backward induction in modified entry deterrence game

bold branches in Fig. 5.8a. In this game, backward induction can be used to compute sequentially rational choices at each node, as shown in Fig. 5.8b, c. ◁

In Example 5.8, sequentially rational actions at each node constitute pure strategy Nash equilibrium ((DE, L), F) of the game. Zermelo's Theorem Zermelo (1913) (Theorem 5.1) provides the assurance that, for any finite extensive-form game of perfect information, pure strategy Nash equilibrium always exists. Furthermore, pure strategy Nash equilibrium can be computed using backward induction. At Node II, Firm $\mathscr{I}$ can correctly infer that Firm $\mathscr{E}$ will always choose L at Node III, and the best response of Firm $\mathscr{I}$ is to choose F at Node II. The best response of Firm $\mathscr{E}$, to the choice of F of Firm $\mathscr{I}$, is DE at Node I. It can be verified that unilateral deviations from these strategies are suboptimal for each firm.

Theorem 5.1 (Zermelo's Theorem) *Every finite extensive-form game of perfect information has a pure strategy Nash equilibrium.* ◀

Mixed strategies in extensive-form games are used to define reasonable solution concepts for the extensive-form games of imperfect information. Players can randomize over the available pure strategies similar to mixed strategies for normal-form games (Sect. 3.2).

Definition 5.2 (*Mixed Strategy in Extensive-Form Games*) A mixed strategy of player i in extensive-form games is a probability distribution over all the pure strategies available to the player. ◀

In Example 5.8, The four possible strategies for Firm $\mathscr{E}$ are

$$s_{\mathscr{E}}^1 = (\text{E,R}); \quad s_{\mathscr{E}}^2 = (\text{E, L}); \quad s_{\mathscr{E}}^3 = (\text{DE, R}); \quad s_{\mathscr{E}}^4 = (\text{DE, L})$$

A mixed strategy of Firm $\mathscr{E}$ is $(\frac{1}{4}s_{\mathscr{E}}^1, \frac{1}{4}s_{\mathscr{E}}^2, \frac{1}{4}s_{\mathscr{E}}^3, \frac{1}{4}s_{\mathscr{E}}^4)$ in which each possible pure strategy is chosen with probability $\frac{1}{4}$.

5.2 Strategies in Extensive-Form Games

In extensive-form games, rather than considering a probability distribution over all possible pure strategies, a player can randomize over the possible actions available at each information set. This way of randomization forms a *behavior strategy*.

Definition 5.3 (*Behavior Strategy in Extensive-Form Games*) A behavior strategy of player i assigns a probability to each action in A_i^K for each information set $K \in I_i$, where I_i is the *set* of information sets of player i and $K \in I_i$ is an information set with A_i^K is the action set at K. ◂

Kuhn's theorem Kuhn (1953) proves that for extensive-form games with perfect recall,[5] mixed and behavior strategies are equivalent—they lead to the same probability distribution for outcomes. Example 5.9 illustrates Kuhn's theorem. We denote a mixed (or behavior strategy) of player i in extensive-form games as σ_i, and the set of such strategies as Σ_i.

Theorem 5.2 (*Kuhn's Theorem*) *For extensive form games with perfect recall, every mixed strategy has an equivalent corresponding behavior strategy, and every behavior strategy has at least one equivalent mixed strategy.* ◂

Example 5.9 (*Ultimatum Game*) Consider a variant of the famous *Ultimatum Game*[6] with two distinct possible actions available to the proposer (rather than a continuum of actions as in Example 5.4. The proposer can choose to be Greedy (G) or Fair (F). After observing the action of the proposer, the responder can choose to either accept or reject the proposed split. Figures 5.9 and 5.10 show extensive-form and normal-form representations of the game, respectively.

Using Definition 5.2, a mixed strategy of the responder is a probability distribution over all his possible pure strategies—Aa, Ar, Ra, Rr. The first letter represents the action recommended at Node I that is reached if the proposer chooses G, while the second letter represents the action recommended at Node II that is reached if the proposer chooses F.

Assume that the responder chooses the following mixed strategy: ($\frac{1}{3}$Aa, $\frac{1}{3}$Ar, $\frac{1}{3}$Ra, 0Rr). Given this mixed strategy, we can construct the joint probability distribution table as shown in Fig. 5.11a.

---

[5] A game has *perfect recall* if the players are not *forgetful*—they remember what they did earlier or knew earlier in the game. The following extensive-form game does not have perfect recall as Player $\mathscr{E}$ cannot distinguish between nodes **x** and **x**', which means the player forgot the earlier choice of DE.

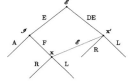

[6] The Ultimatum Game is an asymmetric sequential two-player game with the proposer who offers to split a dollar. If the responder accepts the offer, the game stops with the offered split; otherwise, the game stops with neither player receiving anything. Schuster (2017) proposes an interesting solution concept for the Ultimatum Game that leads to the split, which follows the *golden ratio*.

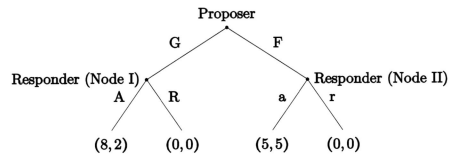

Fig. 5.9 Ultimatum game in extensive form

		Responder			
		Aa	Ar	Ra	Rr
Proposer	G	8,2	8,2	0,0	0,0
	F	5,5	0,0	5,5	0,0

Fig. 5.10 Ultimatum game in normal-form

	A	R	
a	1/3	1/3	2/3
r	1/3	0	1/3
	2/3	1/3	

(a) Case I

	A	R	
a	5/12	1/4	2/3
r	1/4	1/12	1/3
	2/3	1/3	

(b) Case II

Fig. 5.11 Joint probability tables

If the proposer chooses G, Node I is reached, at which the responder has two possible actions, A or R. The responder can randomize between these actions. Using Fig. 5.11a, the responder chooses A with probability 2/3 and R with probability 1/3 at Node I. If the proposer chooses F, the responder chooses a with probability 2/3 and r with probability 1/3. This is the *equivalent corresponding behavior strategy* of the responder for the given mixed strategy.

Figure 5.11b shows the equivalent corresponding behavior strategy for the mixed strategy $(\frac{5}{12}Aa, \frac{1}{4}Ar, \frac{1}{4}Ra, \frac{1}{12}Rr)$. The corresponding behavior strategies associated with the mixed strategies are identical. In other words, a behavior strategy can have more than one equivalent mixed strategy, which is the second part of Kuhn's theorem. ◁

5.3 Solution Concepts for Games in Extensive Form

5.3.1 *Subgame Perfect Nash Equilibrium (SPNE)*

As mentioned earlier, the concept of sequential rationality is essential for defining reasonable solution concepts for extensive-form games. In this section, we discuss the connections between sequential rationality and solution concepts for extensive-form games. Let us start with examples.

Example 5.10 (*Entry Deterrence Game Revisited*) In the entry deterrence game in Example 5.1, a reasonable outcome is (E, A). This is one pure strategy Nash equilibrium in the game. Are there any more pure strategy Nash equilibrium in the game? The unique normal-form representation of the game is shown in Fig. 5.12. It can be verified that the game has two pure strategy Nash equilibria—(E, A) and (DE, F).

Using backward induction, the sequentially rational choices in the game are shown as bold branches in Fig. 5.13. The choice of action DE by Firm $\mathscr{E}$ at Node I can be sequentially rational if it has a belief that Firm $\mathscr{I}$ will choose F at Node II if Firm $\mathscr{E}$ chooses E. However, it is not reasonable for Firm $\mathscr{E}$ to have such a belief as the choice of F by Firm $\mathscr{I}$ at Node II is not sequentially rational as shown in Fig. 5.13.

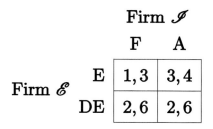

Fig. 5.12 Entry deterrence game in normal form

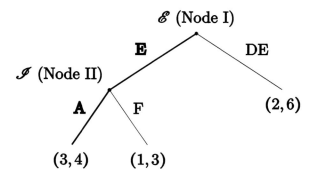

Fig. 5.13 Sequentially rational choices in entry deterrence game

Nash equilibrium (DE, F) is not reasonable as it involves sequentially irrational choices. (E, A) is the only reasonable Nash equilibrium in this game. A reasonable solution concept should satisfy the requirements of sequential rationality.

It is important to note the connections between beliefs and sequential rationality, which is explored further in Sect. 5.3.2. ◁

Example 5.11 (*Modified Entry Deterrence Game Revisited*) In Example 5.8, backward induction leads to a pure strategy Nash equilibrium is ((DE, L), F). This Nash equilibrium involves sequentially rational actions at each of the decision nodes.

The game has two other pure strategies Nash equilibria ((E, R), A) and ((DE, R), F), which can be computed using normal-form representation (Fig. 5.14). Both of them are not reasonable as they involve sequentially irrational choices at some of the decision nodes.

$$\text{Firm } \mathscr{I}$$

		F	A
	E, R	$\frac{1}{2}, 2$	$3, 4$
Firm $\mathscr{E}$	**E, L**	$1, 5$	$3, 4$
	DE, R	$2, 6$	$2, 6$
	DE, L	$2, 6$	$2, 6$

Fig. 5.14 Modified entry deterrence game in normal form ◁

The concept of sequential rationality refines the set of Nash equilibria. For extensive-form games, one of the refinements of Nash equilibrium that captures sequential rationality is *Subgame Perfect Nash Equilibrium* (SPNE).[7] SPNE needs the definition of subgames (Definition 5.4).

Definition 5.4 (*Subgame*) A decision node X in an extensive-form game initiates a subgame if the following set of conditions is met:

- It must initiate from an information set containing a single decision node and contain all the decision nodes succeeding this node and only these nodes.
- In case of multiple decision nodes in an information set, all these nodes must be part of the same subgame. ◀

The entry deterrence game from Example 5.1 has two subgames—one subgame starts at the decision node of Firm $\mathscr{I}$ (shown in red box in Fig. 5.15), and the other is the whole game (shown in the blue box in Fig. 5.15).

---

[7] Selten (1965) proposes SPNE as a solution concept for extensive-form games. Reinhard Selten shared the 1994 Nobel Memorial Prize in Economic Sciences with John Nash and John Harsanyi.

5.3 Solution Concepts for Games in Extensive Form

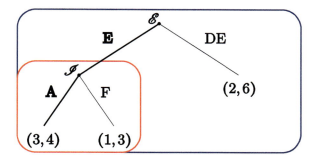

Fig. 5.15 Subgames in entry deterrence game

Example 5.12 (*Centipede Game*) Consider the Centipede game first introduced by Rosenthal (1981), where a player has the option of either keeping a smaller share of money or passing it on to the next player who has the same options but with a larger share. The pot keeps growing until the terminal node is reached. The Centipede game is shown in Fig. 5.16. This is extensive-form game of perfect information, and a subgame starts at each decision node. The Centipede game in Fig. 5.16 has four subgames, as shown in Fig. 5.17.

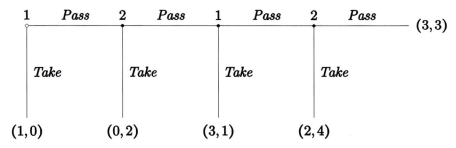

Fig. 5.16 Centipede game

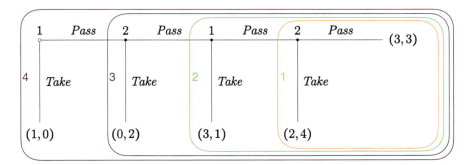

Fig. 5.17 Subgames in centipede game

With the background on sequential rationality and subgames, we now define subgame perfect Nash equilibrium (Definition 5.5).

Definition 5.5 (*Subgame Perfect Nash Equilibrium*) In an extensive-form game with n players, a strategy profile $\sigma = (\sigma_1, \sigma_2, \ldots, \sigma_n)$ is subgame perfect Nash equilibrium (SPNE) if it induces Nash equilibrium in every subgame of the extensive-form game. ◀

Any strategy that is not sequentially rational cannot be part of SPNE, and hence, SPNE is a reasonable solution concept for extensive-form games. It is important to note that every SPNE is a Nash Equilibrium, but the converse is not true. Furthermore, in a finite extensive-form game of perfect information, subgame perfect Nash equilibria are pure strategy Nash equilibria that are computed using backward induction. We know from Zermelo's theorem (Theorem 5.1) that a finite extensive-form game of perfect information has a pure strategy Nash equilibrium, and hence, *a finite extensive-form game of perfect information has a pure strategy subgame perfect Nash equilibrium, which is also unique if no player has same payoffs at any two terminal nodes.*

The process of backward induction can also be used to compute SPNE in extensive-form games of imperfect information as illustrated in Example 5.13.

Example 5.13 Consider the game in Fig. 5.18. The game has two subgames—the first subgame is the whole game itself (shown in the blue box), while the second subgame starts when P_1 chooses E (shown in the red box).

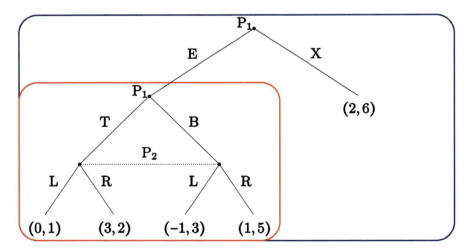

Fig. 5.18 SPNE in game of imperfect information

The subgame in the red box is a game of imperfect information. In this subgame, action R strictly dominates L for P_2. P_1 at the starting node of this subgame can reason out that P_2 will choose R at his decision node, and hence, it is optimal for P_1

5.3 Solution Concepts for Games in Extensive Form

to choose T. Hence, P_2 has a belief that he is at the left node of his information set with probability equal to one. (T, R) is the unique Nash equilibrium in this subgame. The unique SPNE in the entire game is (ET, R), which can be computed using backward induction.

We can verify that the entire game has two other pure strategy Nash Equilibria—(XT, L) and (XT, R). However, they do not induce Nash equilibrium in the subgames; hence, they are not SPNE. ◁

5.3.2 Sequential Equilibrium

Although SPNE is a reasonable solution for extensive-form games, its applicability is limited for extensive-form games with no strict subgames[8] as discussed in Example 5.14. This example is motivated from Mas-Colell et al. (1995).

Example 5.14 (*Entry Deterrence with Different Entry Strategies*) Consider a variation of the entry deterrence game (Example 5.1). Firm $\mathscr{E}$ can enter the market using one of the two actions—E_1 if it plans to sufficiently differentiate its market offering to target a niche set of customers and charge high prices, or E_2 if it plans to attract customers based on extremely competitive pricing. Firm $\mathscr{I}$ is unaware of the choice of E_1 or E_2 and can choose to accommodate (A) or fight (F) The game is represented in Fig. 5.19. The game has two pure strategy Nash equilibria (DE, F) and (E_1, A). As the game has no strict subgames, both pure strategy Nash equilibria are also SPNE. (DE, F) is not reasonable as it involves the sequentially irrational choice of F by Firm $\mathscr{I}$ at its information set. Choosing A is sequentially rational for Firm $\mathscr{I}$. In this example, SPNE involves sequentially irrational strategies that lead to unreasonable outcomes.

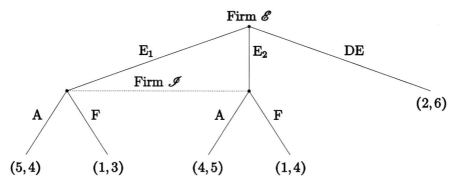

Fig. 5.19 Entry deterrence game with different entry strategies ◁

---

[8] Whole games are not strict subgames. In Fig. 5.18, the subgame in the red box is the only strict subgame.

Can SPNE be refined to provide reasonable outcomes in extensive-form games with no strict subgames? *Sequential equilibrium* proposed by Kreps and Wilson (1982) is one such refinement. Sequential equilibrium has two components

(i) a strategy profile σ that prescribes behavior strategy at each information set for each player.
(ii) a system of beliefs μ assigned to every information set K is a probability distribution over the nodes in K. If player i's information set $K \in I_i$ has been reached, μ is the belief of player i about its position in the information set, and it should satisfy the following condition

$$\sum_{x \in K} \mu(x) = 1$$

Sequential equilibrium is a refinement of SPNE that removes unreasonable equilibrium like (DE, F) in Example 5.14; hence, the strategies that constitute sequential equilibrium should be sequential rational. Given beliefs μ_K at the information set K of player i, σ_i is sequentially rational at K if σ_i is the best response of player i to the strategy profile σ_{-i} used by other players from the information set K onwards in the game. Furthermore, beliefs μ_K at information set K are computed using strategies involved in reaching K, and the probability at each node $x \in K$ is updated using Bayes' rule

$$\mu(x) = \frac{\mathsf{P}(x|\sigma)}{\sum_{x' \in K} \mathsf{P}(x'|\sigma)}$$

In Example 5.14, if Firm $\mathscr{E}$ uses behavior strategy $\sigma_{\mathscr{E}} = (\frac{1}{3}, \frac{1}{3}, \frac{1}{3})$ as shown in Fig. 5.20, then $\mathsf{P}(x|\sigma) = \mathsf{P}(x'|\sigma) = \frac{1}{3}$. Using Bayes' rule

$$\mu(x) = \mu(x') = \frac{1}{2}$$

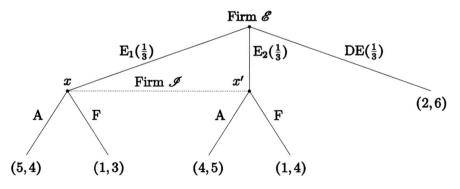

Fig. 5.20 Computing beliefs using strategies

5.3 Solution Concepts for Games in Extensive Form

Beliefs of Firm $\mathscr{I}$ are computed using behavior strategy $\sigma_{\mathscr{E}} = (\frac{1}{3}, \frac{1}{3}, \frac{1}{3})$. This behavior strategy is not sequentially rational for Firm $\mathscr{E}$. Given $\mu(x) = \mu(x') = \frac{1}{2}$, Firm $\mathscr{I}$ chooses A (A strictly dominates F for Firm $\mathscr{I}$), and then the best response of Firm $\mathscr{E}$ is $\sigma_{\mathscr{E}} = (1, 0, 0)$. Recall (E_1, A) is one of the pure strategy Nash equilibrium. It means a strategy profile and a system of beliefs should be *consistent* in sequential equilibrium. Sequential equilibrium uses a stronger notion of consistency based on strictly mixed behavior strategy profiles. In a *strictly mixed behavior strategy profile*, every action at every information set is chosen with positive probability. Using a strictly mixed behavior strategy profile, every information set is reached with a positive probability, and the beliefs at each information set can be computed using Bayes' rule. Now, we define consistency in sequential equilibrium (Definition 5.6).

Definition 5.6 (*Consistency in Sequential Equilibrium*) A strategy profile σ and a system of beliefs μ are consistent if there is a sequence of strictly mixed behavioral strategies $\{\sigma^k\}_{k=1}^{\infty}$, with $\lim_{k \to \infty} \sigma^k = \sigma$, such that $\mu = \lim_{k \to \infty} \mu^k$, where μ^k are strictly positive beliefs derived from strategy profile σ^k using Bayes' rule. ◀

Definition 5.7 (*Sequential Equilibrium*) A strategy profile and system of beliefs (σ, μ) is a sequential equilibrium in an extensive-form game if they are consistent as in Definition 5.6 and σ is sequentially rational at each information set. ◀

In Definition 5.6, σ^k are strictly mixed behavioral strategy profiles that are close to equilibrium strategy profile σ. They ensure that each information set is reached with positive probability. Consistency in Definition 5.6 means that beliefs μ^k at any information are not arbitrary. They are justifiable as they are consistent with σ^k. We elaborate further on consistency and sequential rationality in Example 5.15. This example is motivated from Osborne and Rubinstein (1994).

Example 5.15 Consider the extensive-form game shown in Fig. 5.21. (R, r) is Nash equilibrium in this game. As the game has no subgames, (R, r) is also SPNE.

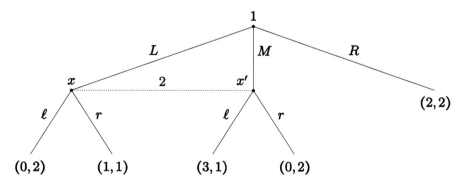

Fig. 5.21 Consistency and sequential rationality

Why does player 2 choose r? r is sequentially rational if player 2 believes that he is at node x' with probability $\mu(x') \geq \frac{1}{2}$.[9] However, in Nash equilibrium (R, r), player 2's information set is not reached, and beliefs, which make r sequentially rational choice, are not consistent with the strategy of player 1.

To ensure consistency of beliefs and strategy profile, we assume player 1 uses a strictly mixed behavior strategy $\sigma_1 = ((1-k)\epsilon, k\epsilon, 1-\epsilon)$. $\epsilon > 0$ is small perturbation and hence σ_1 is close to equilibrium strategy R of player 1. For player 2, $P(x|\sigma_1) = (1-k)\epsilon$ and $P(x'|\sigma_1) = k\epsilon$. Using Bayes' rule

$$\mu(x) = 1-k \text{ and } \mu(x') = k$$

For $k \geq \frac{1}{2}$, r is a sequentially rational strategy for player 2. ◁

Example 5.16 (*Entry Deterrence Game with Incomplete Information Revisited*) Let us revisit the entry deterrence game with incomplete information from Example 5.6. Extensive-form representation is shown in Fig. 5.22. This is a game of incomplete information in which Firm $\mathscr{I}$ has asymmetric information about the set-up cost of Firm $\mathscr{I}$. As in Bayesian games (Sect. 3.4.5), the game has three players—Firm $\mathscr{I}$, Firm $\mathscr{E}_H$, and Firm $\mathscr{E}_L$.

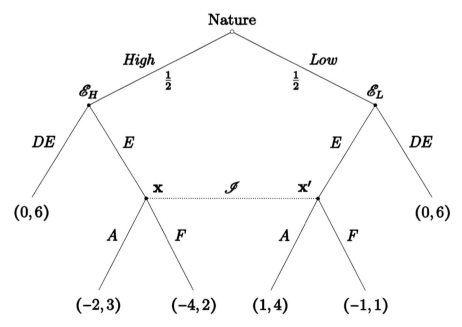

Fig. 5.22 Entry deterrence game with incomplete information

---

[9] For player 2, $r \succ \ell$ if $\mu(x) + 2\mu(x') > 2\mu(x) + \mu(x')$, and $\mu(x) + \mu(x') = 1$.

In this game, A strictly dominates F for Firm $\mathscr{I}$, regardless of Firm $\mathscr{I}$'s beliefs about its position $\mathbf{x}$ or $\mathbf{x}'$. Firm $\mathscr{E}_H$ and Firm $\mathscr{E}_L$ can reason it out that Firm $\mathscr{I}$ will always choose A, if given a chance to choose.

If Firm $\mathscr{I}$ always chooses A, then the best response of Firm $\mathscr{E}_H$ is DE, and the best response of Firm $\mathscr{E}_L$ is E. Therefore, if Firm $\mathscr{I}$ is given an opportunity to move, it can reason that it is node $\mathbf{x}'$. Firm $\mathscr{I}$ assigns beliefs $\mu(\mathbf{x}) = 0$ and $\mu(\mathbf{x}') = 1$, which are consistent with the strategies of Firm $\mathscr{E}_H$ and Firm $\mathscr{E}_L$. With these beliefs, A is still sequentially rational for Firm $\mathscr{I}$, and hence DE and E continue to be the best responses of Firm $\mathscr{E}_H$ and Firm $\mathscr{E}_L$, respectively. *It means strategy profile* $(\sigma_{\mathscr{E}_H} = DE, \sigma_{\mathscr{E}_L} = E, \sigma_{\mathscr{I}} = A)$ *with system of beliefs* $\mu(\mathbf{x}) = 0$ *and* $\mu(\mathbf{x}') = 1$ *constitute sequential equilibrium as the strategy profile and the system of beliefs are consistent and strategy profile is sequentially rational.* ◁

Problems

Problem 5.1 Consider the finite-horizon bargaining game in Example 5.4. Instead of discounting, each player pays a cost of ₹5 to advance from one period to the next. If they do not agree at the end of three periods, both get nothing. Compute subgame perfect Nash equilibria (SPNE). ◀

Problem 5.2 In 2004, the Nobel Prize in Economics was awarded to Finn Kydland and Edward Prescott. Their model provided insights into the importance of creating independent central banks like the Reserve Bank of India. In their model, at $t = 0$, an investor[10] I decide whether or not to invest in a country. At $t = 1$, the central bank CB decides on the interest rates that may lead to high inflation (H) or low inflation (L). The payoff structure is depicted in the following figure:

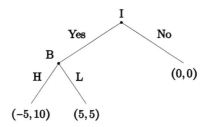

i Identify Nash equilibria in the above game.
ii Identify subgame perfect Nash equilibria (SPNE) in the above game. ◀

Problem 5.3 Consider the finite-horizon bargaining game in Example 5.4. There are infinite rounds of bargaining. Player 1 makes an offer, which Player 2 can accept

---
[10] Adapted from Brinkhuis and Tikhomirov (2005).

104   5 Games in Extensive Form

or reject. Then, Player 2 makes a counteroffer, which Player 1 can accept or reject, and so on.

  i Model this as an extensive form game.
  ii Identify the unique subgame perfect Nash equilibrium (SPNE) of this game. ◄

Problem 5.4 Compute sequential equilibria of the entry deterrence game with different entry strategies (Example 5.14). ◄

References

Brinkhuis, J., Tikhomirov, V. M. (2005). *Optimization: Insights and applications.*
Kreps, D. M., & Wilson, R. (1982). Sequential equilibria. *Econometrica: Journal of the Econometric Society, 863–894.*
Kuhn[1], H. (1953). Extensive games and the problem of information. *Contributions to the Theory of Games, 24,* 193.
Mas-Colell, A., Whinston, M. D., & Green, J. R. (1995). *Microeconomic theory.* New York: Oxford University Press.
Osborne, M. J., & Rubinstein, A. (1994). *A course in game theory.* Cambridge, Mass: MIT Press.
Rosenthal, R. W. (1981). Games of perfect information, predatory pricing and the chain-store paradox. *Journal of Economic theory, 25*(1), 92–100.
Rubinstein, A. (1982). Perfect equilibrium in a bargaining model. *Econometrica, 50*(1), 97–109.
Schuster, S. (2017). A new solution concept for the ultimatum game leading to the golden ratio. *Scientific Reports, 7*(1), 5642.
Selten, R. (1965). SPIELTHEORETISCHE BEHANDLUNG EINES OLIGOPOLMODELLS MIT NACHFRAGETRÄGHEIT: TEIL I: BESTIMMUNG DES DYNAMISCHEN PREISGLE-ICHGEWICHTS. *Zeitschrift für die gesamte Staatswissenschaft/Journal of Institutional and Theoretical Economics, 121*(2), 301–324.
Ståhl, I. (1972). *Bargaining theory.* Stockholm: Economics Research unit.
von Stackelberg, H. (1934). *Marktform und Gleichgewicht.* J. Springer.
Zermelo, E. (1913). Über eine anwendung der mengenlehre auf die theorie des schachspiels. In *Proceedings of the fifth international congress of mathematicians* (vol. 2, pp. 501–504). Cambridge: Cambridge University Press.

Chapter 6
Games in Extensive Form: Applications in OM

In Chap. 5, we studied games in extensive form. This chapter discusses important operations management (OM) applications[1] that can be represented as extensive-form games.

6.1 Capacity Decisions[2]

Capacity decisions play a key role in the way in which a firm decides to fulfill the market demand for its products. Before establishing a capacity, a firm must have complete clarity over its business strategy. It must determine the type of market, whether new or existing, it plans to target. Further, questions like the decisions to outsource versus the decision to build in-house must be resolved beforehand. Only then can the issues of the capacity size be settled. Anticipated demand is a good indicator for deciding the capacity size. If the demand for the product is very high in the near future, it may not be able to fulfill the demand completely. Hence, it would make sense for the firm to outsource its requirements to capable suppliers. A supplier's capacity decisions also follow a similar logic. If he expects a windfall of orders for a product in the medium to long term, he may end up investing in highly specialized equipment, leading to low variable costs per unit. On the other hand, he may be reluctant to opt for specialized capacity if the product's demand is likely to drop or the product/manufacturing technology is susceptible to becoming obsolete soon. Due to this fear of obsolescence, a manufacturer may thus be willing to set up partly *specialized* and *fungible* capacities. Figure 6.1 illustrates the nature of specialized and fungible capacities. Song et al. (2020) and Van Mieghem (2003) provide a comprehensive review of strategic capacity.

---

[1] It is important to note that the notations used in each section in this chapter are specific to that application.

[2] In this section, the OM actions are on the *asset side* in the VCAP framework.

© The Author(s), under exclusive license to Springer Nature Singapore Pte Ltd. 2024
R. K. Amit, *Game Theory with Applications in Operations Management*, Springer Texts in Business and Economics, https://doi.org/10.1007/978-981-99-4833-8_6

Specialized Capacity   Fungible Capacity

Fig. 6.1 Nature of specialized and fungible capacities

The manufacturing of COVID-19 vaccines presents an interesting case for analyzing capacity decisions. Traditionally, a vaccine requires upwards of 10 years for development and regulatory approval. But, as observed in the case of COVID-19, developers achieved regulatory approval of their vaccine candidates less than a year into the pandemic. With vaccines having enormous value in a pandemic, bridging the gap between vaccine approval and supply to the market became important. For this purpose, vaccine developers planned to set up capacity before their candidate attained regulatory approval. The developers faced two choices: to set up manufacturing capacity in-house or to outsource it to contract manufacturers. Ideally, the contractors would set up capacity only if there was a guarantee of success in the trials. But if the candidate did succeed, the manufacturers would be in line for huge profits. To sweeten the pot, the developers can propose to contribute to a manufacturer's cost of setting up capacity. To avoid the scenario of the investment being completely sunk in case of failure in trials, contractors plan for partially flexible capacity. For example, if vaccine candidate V fails to get regulatory approval, doses for some other vaccine candidate V' can be manufactured using the same facilities at some additional cost.

To make the scene more realistic, we consider a case where the vaccine is being developed using a novel technique requiring the capacity to be built up from scratch. A similar scenario was observed for manufacturing *mRNA vaccines* for COVID-19. For a given developer demand, a contract manufacturer may choose to invest in an at-risk specialized capacity, or fungible capacity. Setting up at-risk capacity allows for production to start instantly, while a fungible capacity may require some time to commence production, leading to supply shortages. The investment in at-risk capacity is made before the vaccine trial results are declared. The fungible capacity, however, can be reconfigured for producing the vaccine candidate after the vaccine is approved for administration. This presents a misalignment of incentives faced by supply chains, as discussed in Sect. 4.4. The developer wants the manufacturer to invest in the at-risk capacity to meet the supply requirements without lag, while the manufacturer may wish to await the regulatory decision for the vaccine candidate. Game theory provides a solution to this misalignment problem, leading to an increased supply of doses to the market in a shortened time frame.

Sun et al. (2023) study this lack of coordination between the developer and the manufacturer. To share the burden of risk with the manufacturer, the developer decides to contribute towards setting up the at-risk capacity. The manufacturer takes his decision of capacity set-up on the *signals* he receives from the developer. With the information asymmetry regarding the probability of the vaccine candidate's suc-

6.1 Capacity Decisions

cess, the manufacturer needs some credible signals from the developer. There may be some non-credible cheap talk (discussed later in Sect. 6.3) from the developer who shares only those reports with the manufacturer that show high chances of her vaccine candidate's success. This goes into the category of *signaling games*. With the manufacturer anticipating only favorable reports from the developer about her vaccine candidate, can there be a proxy for the manufacturer to elicit a truthful response about the vaccine candidate's success? The willingness of the developer to share at-risk capacity costs provides a truthful signal to the manufacturer about the vaccine candidate's success.

This situation has sequential moves of the developer and the manufacturer. The vaccine developer with vaccine candidate V chooses her share of at-risk capacity investment. Observing the developer's decision, the manufacturer determines his at-risk capacity decision. Consider x to be the proportion of cost that the developer agrees to share, and let y be the ratio of *at-risk* capacity set up by the manufacturer as a proportion of the total capacity q. At-risk capacity κ_R and fungible capacity κ_F of the manufacturer are

$$\kappa_R(y) = yq$$
$$\kappa_F(y) = (1 - y)q$$

Due to compressed timelines, new technologies, and an uncertain labor force, manufacturing COVID-19 vaccines can be assumed to follow *diseconomies of scale*. The at-risk capacity cost function can be taken as $\Gamma_R(y) = \gamma(\kappa_R(y))^2$, where γ is a positive coefficient. The fungible capacity is set up only after the vaccine gets the approval. The same function can be used to quantify the manufacturing cost of the fungible capacity. While the at-risk capacity is used for manufacturing the vaccine for the entire duration, the fungible capacity has a lead time $l \in (0, 1)$ as a ratio of the total production phase. At the end of the production phase, the total quantity produced by the manufacturer is $\kappa_V = \kappa_R(y) + (1 - l)\kappa_F(y)$.

In case vaccine V fails, the manufacturer has the opportunity to produce doses for the developer with vaccine candidate V'. We assume that the developer with vaccine candidate V' uses a different technology than developer with vaccine candidate V, and the manufacturer needs to alter the manufacturing process at an additional cost. A coefficient $\theta > 0$ is defined that captures this relationship. We assume that the manufacturer can avail of an order for the entire quantity capacity q from the developer with vaccine candidate V'. The at-risk production cost function for the developer with vaccine candidate V' is $\Gamma'_R = \theta\gamma(\kappa_R(y))^2$ and the fungible production cost function is $\Gamma'_F = \theta\gamma(\kappa_F(y))^2$. To produce vaccine V', the manufacturer takes a lead time $l_B \in (0, 1)$. So the quantity produced is $\kappa_{V'} = (1 - l_B)q$.

The situation is represented as an extensive-form game in Fig. 6.2. Let c_V and $c_{V'}$ be the per unit manufacturing costs, and w_V and $w_{V'}$ be the per unit price charged by the manufacturer to the developer with vaccine V and to the developer with vaccine V', respectively. Let p be the unit selling price for the developer with vaccine V. If

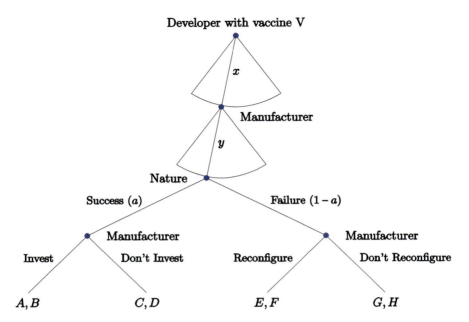

Fig. 6.2 Signaling game between the developer of vaccine V and the manufacturer

Table 6.1 Payoffs in the signaling game

A	$a(-x\gamma\kappa_R^2 + (p - w_V)(\kappa_R + (1-l)\kappa_F) + (q - \kappa_R + (1-l)\kappa_F)s)$
B	$a(x\gamma\kappa_R^2 - \gamma\kappa_R^2 + (w_V - c_V)(\kappa_R + (1-l)\kappa_F) - \gamma\kappa_F^2 - (q - \kappa_R - (1-l)\kappa_F)s)$
C	$a(-x\gamma\kappa_R^2 + (p - w_V)\kappa_R + (q - \kappa_R)s)$
D	$a(x\gamma\kappa_F^2 - \gamma\kappa_R^2 + (w_V - c_V)\kappa_R - (q - \kappa_R)s)$
E	$-(1-a)x\gamma\kappa_R^2$
F	$(1-a)(x\gamma\kappa_R^2 - \gamma\kappa_R^2 + ((1-l_B)q(w_{V'} - c_{V'}) - \theta\gamma\kappa_R^2 - \theta\gamma\kappa_F^2)$
G	$-(1-a)x\gamma\kappa_R^2$
H	$(1-a)(x\gamma\kappa_R^2 - \gamma\kappa_R^2)$

the manufacturer fails to meet the order q, a penalty of s per unit is charged to the manufacturer. The payoffs in Fig. 6.2 are shown in Table 6.1.

This is an extensive-form game with perfect information, and subgame perfect Nash equilibrium (SPNE) is the reasonable solution concept (refer Sect. 5.3.1). Given the payoffs in Table 6.1, we can assume that choosing "Invest" and "Reconfigure" brings positive payoffs to the manufacturer. This makes "Invest" dominate "Don't Invest" and "Reconfigure" dominate "Don't Reconfigure" for the manufacturer at the decision nodes after Nature's moves. Using backward induction, the modified game is shown in Fig. 6.3.

6.1 Capacity Decisions

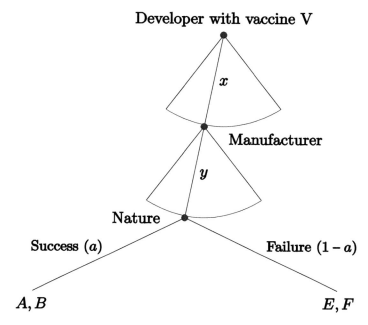

Fig. 6.3 Modified signaling game

In the modified game, the manufacturer determines the at-risk capacity level y after observing the developer's cost-sharing ratio x as a signal. The manufacturer chooses y that maximizes his expected profit, as shown in Eq. 6.1.

$$\max_{y \in [0,1]} \pi_M | x = B + F \tag{6.1}$$

Conjecturing the best response of the manufacturer, the developer determines x that maximizes her expected profit, as shown in Eq. 6.2.

$$\max_{x \in [0,1]} \pi_D = A + E \tag{6.2}$$

Using Eq. 6.1, the manufacturer best response (Eq. 6.3) is

$$y^* = \min\{(\phi_2/(2q\gamma(\phi_1 - x)), 1\} \tag{6.3}$$

where $\phi_1 = 1 + a + 2\theta(1-a)$ and $\phi_2 = al(w_V + s - c_V) + 2q(a + (1-a)\theta)$. The developer is rational to compute the best response function of the manufacturer and then determines her optimal cost-sharing ratio x^*. The modified game has a unique SPNE (x^*, y^*) with x^* and y^* shown in Eqs. 6.4 and 6.5, respectively.

$$x^* = \begin{cases} 0 & \text{if } a \in [0, a_1) \bigcup [a_3, 1] \\ \frac{\phi_1(2al(p-w_V-s)-\phi_2)}{2al(p-w_V-s)+\phi_2} & \text{if } a \in (a_1, a_2) \\ 2q\gamma(1 + (1-a)\theta) - \frac{a(l(w_V+s-c_V))}{2q} & \text{if } a \in [a_2, a_3] \end{cases} \quad (6.4)$$

$$y^* = \begin{cases} al(s + w_V - c_V) + \frac{2q\gamma(a+(1-a)\theta)}{2q\gamma\phi_1} & \text{if } a \in [0, a_1] \\ \frac{2al(p-w_V-s)+\phi_2}{4q\phi_1} & \text{if } a \in (a_1, a_2) \\ 1 & \text{if } a \in [a_2, 1] \end{cases} \quad (6.5)$$

where a_1, a_2, and a_3 are three thresholds of the success probability.

From Eq. 6.3, it can be inferred that the higher value of x incentivizes the manufacturer to set up higher at-risk capacity. Also, as the probability of success a increases, the manufacturer becomes self-inclined to set up a higher proportion of at-risk capacity. SPNE provides wisdom regarding the signals sent by the developer and the subsequent capacity set up by the manufacturer. For low $a \in [0, a_1]$ and high $a \in [a_3, 1]$ values of success probabilities, the developer may not be willing to share any capacity risk—for low a, the developer may not risk losing her investment, while for high a, the manufacturer is self-motivated to set up at-risk capacity. Signaling may incur additional costs even for a competent developer; hence, competent developers may face conditions of insufficient at-risk capacity building. It may be beneficial for highly competent developers to get their credibility assured through third-party regulators.

Continuing with the theme of vaccines, we present another situation. The distribution and uptake of vaccines in the developing world have been a concern for global health organizations for more than two decades. Diseases endemic to developing countries cause a considerable number of avoidable deaths. Pharmaceutical companies are reluctant to manufacture vaccines for Lower and Middle-income Countries (LMICs) owing to the uncertainty in demand and the low purchasing power of those countries. Rather, the available manufacturing capacity is allocated to produce drugs and vaccines for vaccines and drugs with higher expected returns. Kremer and Glennerster (2004) explore the issue of market failures leading to low vaccination rates.

In 2009, the Global Alliance for Vaccines and Immunization (GAVI) proposed to solve this problem using a market-based mechanism, the Advance Market Commitment (AMC). Under this contract, a donor organization (DO) provides a sales subsidy over a tail price paid by the recipient country. In return, the pharmaceutical company (PC) provides a specified quantity of doses. The DO decides the contract parameters, subsidy amount δ, and the quantity cap for $\bar{q}$ eligible for receiving the sales subsidy. Observing the parameters of DO, PC then decides the capacity κ that will be reserved for the production of doses for LMICs. Additionally, DO needs to make these decisions under a fixed donor budget b. Martin et al. (2020) study the different contracts that can be implemented in this setting, taking the AMC implemented for PCV vaccines in 2009 as the base case. We discuss this base case and demonstrate game theory usage to increase vaccination rates in LMICs. Figure 6.4 provides a schematic representation of the sequence of events for the game.

6.1 Capacity Decisions

Fig. 6.4 Sequence of events in AMC

Under the AMC, DO estimates demand $D \in [d_{min}, d_{max}]$ and set contract parameters $\bar{q}$ and δ. PC observes the choices of DO, and decides capacity κ^* that maximizes its payoff π^{PC} in Eq. 6.6.

$$\pi^{PC}(\kappa) = (r-c)\mathbb{E}[\min\{D, \kappa\}] + \delta\mathbb{E}[\min\{D, \kappa, \bar{q}\}] - c_\kappa \kappa \quad (6.6)$$

where r is the tail price per dose paid by the recipient country, c is the per dose manufacturing cost, and c_κ is the per unit capacity commitment costs. κ^* is the best response of PC. The setting is a sequential game that can be represented as an extensive-form game with perfect information with subgame perfect Nash equilibrium (SPNE) as the reasonable solution concept. Using backward induction, DO computes κ^* of PC and determines optimal $\bar{q}^*$ and δ^* to maximize its payoff in π^{DO} in Eq. 6.7. It is assumed that DO needs to meet the budgetary constraint.

$$\max_{\delta, \bar{q}} \pi^{DO} = \mathbb{E}[\min\{D, \kappa^*(\delta, \bar{q})\}]$$
$$\text{Subject to:} \quad \delta\mathbb{E}[\min\{D, \kappa^*(\delta, \bar{q}), \bar{q}\}] \leq b \quad (6.7)$$

Kremer et al. (2020) review the success of the contract used for the purchase of pneumococcal conjugate vaccine (PCV) for LMICs. Donors pledged a total of $1.5 billion towards this pilot AMC. GAVI introduced a price cap per dose and invited tenders from pharmaceutical companies to supply the vaccine doses. In 2010, the first tender was set for 60 million doses. The initial price finalized by GAVI was $3.5 per dose with $0.2 as the co-payment and an average subsidy level of $0.75. Subsequent tenders were issued later, with the manufacturers increasing their capacity to fulfill the high demand for doses. The success of the AMC for PCV can be verified by comparing it with another GAVI-supported vaccination program. The rotavirus vaccination program did not involve using an AMC and took 5 years longer to reach the coverage levels for PCV.

These examples illustrate game theory's use in understanding vaccine supply chain inefficiencies and how they can be mitigated using contract design.

6.2 Leadership in Supply Chains[3]

With advances in technology, each stage in a supply chain has evolved into a highly specialized entity. This has led to the rise of some entities (or stages or agents) holding more power than other entities. Such agents are called *leaders* or *dictators*. For example, brands with high branding power have the capability to keep command of both the upstream and downstream wholesale prices, or a wholesaler having control of a key manufacturing resource can dictate both the upstream and the downstream prices. Such changes in how business is conducted have altered the nature of contracting relationships between the different supply chain agents. Due to double marginalization discussed in Sect. 4.4, competing agents in a supply chain can lower the total supply chain profits (Spengler 1950). An interesting question arises as to how the presence of a leader in a supply chain changes the profit levels of the upstream and downstream agents. For example, consider a food supply chain with five agents in Fig. 6.5. The processor in this supply chain exerts an enormous influence. In this case, the processor can dictate the price w at which the food grains are bought from the aggregator. Also, the processor can dictate the selling price p to the wholesaler. This application is motivated by Majumder and Srinivasan (2006), who study the impact of contract leadership on supply chain performance.

Consider a five-agent supply chain akin to the food supply chain in Fig. 6.5, with the manufacturer M as Agent 0 and the retailer R as Agent 4. We analyze the leadership in the supply chain with the help of wholesale-price contracts (refer Sect. 4.4.1). The interior agents of the supply chain are all wholesalers W_1, W_2, and W_3, with W_2 as the leader in the supply chain. Furthermore, each agent i, except the manufacturer, incurs a constant marginal cost of c^i. Figure 6.6 exhibits the flow of information of the dictated wholesale prices and the quantities ordered at that price.

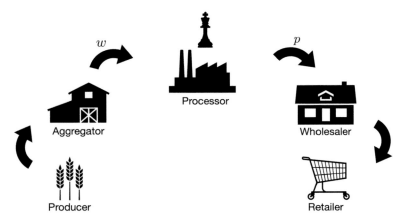

Fig. 6.5 Leadership in food supply chain

---

[3] In this section, the OM actions are on the *process side* in the VCAP framework.

6.2 Leadership in Supply Chains

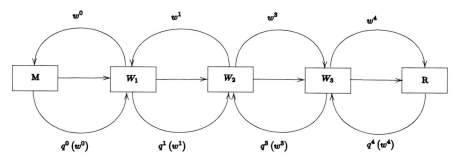

Fig. 6.6 Flow of information

Table 6.2 Dictatorship in supply chain

Node	Dictated	Dictates
M	$w^0, m(q)$	q^0
W_1	w^1	q^1, w^0
W_2 (Leader)	–	w^1, w^3
W_3	w^3	q^3, w^4
R	$w^4, D(p)$	q^4, p

For example, leader W_2 sets wholesale price w^3 for wholesaler W_3, who orders quantity of $q^3(w^3)$ from W_2. Table 6.2 lists all such relationships in the supply chain.

Let us start with solving the problem faced by wholesaler W_3. Leader W_2 dictates the wholesale price w^3 for W_3. Then, W_3 decides the purchase quantity q^3 and the wholesale price w^4 for the downstream agent R. For wholesaler W_3, the quantity q^4 ordered by retailer R depends on the chosen wholesale price w^4. For wholesaler W_3, the profit function is given in Eq. 6.8.

$$\max_{w^4, q^3} \Pi^{W_3} = w^4 \min\{q^3, q^4(w^4)\} - (w^3 + c^3)q^3 \tag{6.8}$$

Further downstream, retailer R sells the goods at price p. The retailer makes the price and quantity decisions, given demand function $D(p)$. The retailer's profit function is given in Eq. 6.9.

$$\max_{p, q^4} \Pi^R = p \cdot \min\{q^4, D(p)\} - (w^4 + c^4)q^4 \tag{6.9}$$

The agents that are upstream of the leader face a similar problem. W_2 dictates the wholesale price w^1 to W_1. Given w^1, W_1 determines the order quantity q^1 and the upstream wholesale price w^0. W_2 must consider the relationship between wholesale price w^0 and the quantity q^0 produced by the manufacturer. For wholesaler W_1, the profit function is given in Eq. 6.10.

$$\max_{q^1, w^0} \Pi^{W_1} = w^1 \min\{q^1, q^0(w^0)\} - (w^0 + c^1)(q^0(w^0)) \tag{6.10}$$

For the given wholesale price w^0, manufacturer M determines the production quantity q^0. The manufacturer's profit function is given in Eq. 6.11.

$$\max_{q^0} \Pi^M = w^0 q^0 - m(q^0) \tag{6.11}$$

where $m(q)$ is the total cost function of manufacturer M.

Leader W_2 decides the downstream wholesale price w^3 and also the upstream wholesale price w^1. For W_2, the profit function is given in Eq. 6.12.

$$\max_{w^1, w^3} \Pi^{W_2} = w^3 \min\{q^1(w^1), q^3(w^3)\} - (w^1 + c^2)(q^1(w^1)) \tag{6.12}$$

The setting can be modeled as a perfect-information extensive-form game with subgame perfect Nash equilibrium (SPNE) as the reasonable solution concept. SPNE can be computed using the process of backward induction. W_3 and R are the downstream agents of W_2. The problem for R is solved first. Retailer R computes his best response to the wholesale price w^4, which is decided by W_3. The problem is then solved for wholesaler W_3, who computes his best response to w^3 that is dictated by W_2, and conjectures the best response of R to w^4. Similarly, the problem on the upstream side is solved using backward induction starting from the manufacturer. For wholesale price w^0 decided by W_1, manufacturer M computes his best response. Wholesaler W_1 then computes his best response to w^1 that is dictated by W_2, and conjectures the best response of M to w^0. Conjecturing all the best responses, leader W_3 simultaneously solves the upstream and downstream problems to obtain her decision variables w^1 and w^3.

Consider the case where the total cost function of manufacturer M is given by $m(q) = \alpha_K q^2 + \alpha_q q$ and the demand function $D(p) = A - Bp$, where p is the retail price. The problem can be generalized by denoting the total number of members in the chain as N, and considering i to be the position of the leader. On solving the above game, equilibrium quantity and profit function are

$$q^*_{(N,i)} = \frac{A - B\bar{c}}{2^i B\alpha_K + 2^{N-i+1}}$$

$$\Pi^k_{N,i} = \begin{cases} 2^{k-1}\alpha_K \chi & \forall k \in \{1 \ldots i-1\} \\ \left(2^{i-1}\alpha_K + \dfrac{2^{N-1}}{B}\right)\chi & \text{for } k = i \\ \dfrac{2^{N-k}}{B}\chi & \forall k \in \{i+1 \ldots N\} \end{cases}$$

where $\chi = \dfrac{(A - B\bar{c})^2}{(2^i B\alpha_K + 2^{N-i+1})^2}$ and $\bar{c} = \alpha_q + \sum_{j=1}^{N} c^j$.

6.3 Cheap Talk in Operations Management

Now, consider the situation when the supply chain is managed by a centralized decision-maker. The objective of the centralized decision-maker is to maximize the overall supply chain profits, given in Eq. 6.13.

$$\max_{p,q} \quad p \cdot \min\{D(p), q\} - \left(\alpha_K q^2 + \alpha_q q + q \sum_{j=1}^{N} c_j \right) \tag{6.13}$$

Assuming $D(p) = q$, the optimal quantity and optimal total supply chain profit are computed as in Eqs. 6.14 and 6.15, respectively.

$$q_c^* = \frac{A - B\overline{c}}{2 + 2B\alpha_K} \tag{6.14}$$

$$\Pi_C^* = \frac{(A - B\overline{c})^2}{4B(1 + B\alpha_K)} \tag{6.15}$$

It can be verified that the centralized solution has a higher optimal quantity and total profit than the equilibrium solution in the decentralized supply chain. The detrimental effects of competition in a supply chain can be clearly observed. For the decentralized case, profit margins decrease as we move away from the leader. A member closer to the leader obtains double the profit from that of his neighbor. The leader ends up making the highest profits, with her profit equal to twice the sum of her neighbors. This analysis suggests an urgent need for supply chain coordination. Majumder and Srinivasan (2006) discuss coordinating such supply chains using *two-part tariff contracts*.

6.3 Cheap Talk in Operations Management[4]

We discussed cheap talk in Sect. 3.4.4.1, and showed that in Example 2.1, cheap talk by the players is non-credible and hence not self-enforcing. As mentioned in Sect. 3.4.4.1, cheap talk involves costless actions. Each costless action is called a *message*. In game theory, we assume that the players are strategic, and they can use cheap talk to influence the outcome of a game.[5] Although cheap talk does not directly alter the players' payoffs, it can influence the outcome. In some cases, cheap talk being costless is not credible. In such cases, to enhance the credibility of communication,

---

[4] In this section, the OM actions are on the *process side* in the VCAP framework.

[5] An interesting story of cheap talk is from the ancient Indian epic *Mahabharata*. During *Mahayudha* (the Great War) between *Pandavas* and *Kauravas* at Kurushetra, *Yudhishthira*, the Pandava's Chief Commander, was approached by *Drona*, the Kauravas' then-Chief Commander, to inquire about death of his son *Ashwatthama*. Yudhishthira then engaged in cheap talk by confirming the death of Ashwattama, but omitted the contextual part that it was an elephant and not his son. This communication was effective in crippling Drona, and changed the outcome of the war in Pandava's favor.

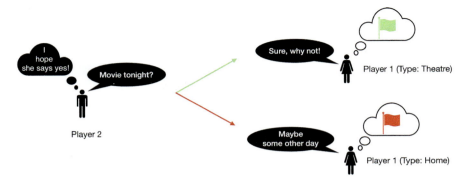

Fig. 6.7 Sequential moves in movie game

a player (the sender) conveys to other players that the talk is not cheap by signaling the cost/effort of the sender before communicating.[6] In Sect. 4.5, proof of work in blockchains is an example of such a cost. In Sect. 6.1, the developer sends a costly signal to the manufacturer by demonstrating her willingness to share the capacity costs. Another classic example is *job market signaling* (Spence, 1978),[7] where a *good* type applicant takes a costly action (education) to differentiate himself from a *bad* type applicant. This section discusses the connections between cheap talk and operations management.

We start the discussion with two illustrative examples of credible and non-credible cheap talk. The first example, called *movie game*, is motivated by Aumann and Hart (2003). Suppose Player 2 wants to ask Player 1 for a movie but does not know if she will be up for it. Player 1 can have two possible types—"Home" and "Theatre". The type can also be interpreted as her mood, whether she wishes to go to the movie. Nature determines this type for her. Regardless of Player 1's type, Player 2 must choose between "Ask" and "Don't Ask". Player 1 does not have a strategic choice as Nature makes the decision for Player 1.[8] Player 2 is not aware of the choice of nature. In this sequential game, Player 2 chooses between "Ask" and "Don't Ask", followed by the response of Player 1 based on her type as shown in Fig. 6.7. The setting has two bimatrix games, one for each type of Player 1, as shown in Figs. 6.8 and 6.9.

Assume that Player 1 is equally likely to be of either type, then Player 2 has a dilemma between "Ask" and "Don't Ask", randomizes over his moves, and gets payoff (0, 0).

---

[6] Games in which one player takes an expensive action to reveal some information about herself (or her *type*) are called *signaling games*. Signals are an approach with which players alleviate the problem of asymmetric information.

[7] For this work, Michael Spence shared the 2001 Nobel Memorial Prize in Economic Sciences with George Akerlof and Joseph E. Stiglitz. We also met Michael Spence in Chap. 1.

[8] Recall Footnote 5 in Sect. 2.2.3, nature is a non-strategic player in game theory.

6.3 Cheap Talk in Operations Management

	Player 2	
	Ask	Don't Ask
	2, 2	−2, −2

Fig. 6.8 Bimatrix game with player 1's type: theater

	Player 2	
	Ask	Don't Ask
	−2, −2	2, 2

Fig. 6.9 Bimatrix game with player 1's type: home

Fig. 6.10 Cheap talk by player 1

To improve the outcome, Player 1 can engage in cheap talk by informing Player 2 of her type, as shown in Fig. 6.10. As the payoffs are aligned for each player, Player 1 has no incentive to lie about her type, as lying may lead to a lower payoff. Player 2 considers Player 1's communication credible and chooses the optimal action. The payoff in this setting is (2, 2). This is an example of *credible cheap talk*.

In the following example, cheap talk is not credible, and the sender engages in costly signals. Consider a situation where candidate C applies for a job and department D is the evaluator. Candidate C has two possible types: "Qualified" and "Under-qualified". These types are the private information of candidate C, and she can engage in cheap talk to inform department D of her type as shown in Fig. 6.11.

The setting has two bimatrix games, one for each type of candidate C, as shown in Figs. 6.12 and 6.13. In this case, department D knows that their interests do not align, and candidate C of type "Under-qualified" has the incentive to engage in *non-credible cheap talk* to signal department D that her type is "Qualified".

As evident from Fig. 6.11, cheap talk by candidate C is non-credible and uninformative. Assume each type of candidate C is equally likely, then department D by choosing "Don't Hire" gets the expected payoff of 1.5, which is higher than the expected payoff of 1.0 by choosing "Hire". It means when candidate C engages in non-credible cheap talk, department D's best response is to ignore all communication from candidate C and choose "Don't Hire". This can be conjectured by candidate C, and she engages in more babbling. This behavior constitutes an equilibrium called *babbling equilibrium*.

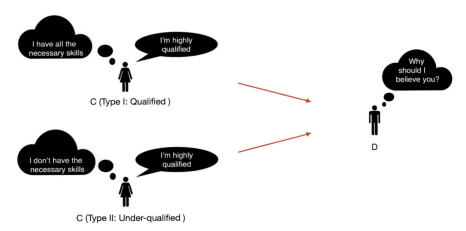

Fig. 6.11 Cheap talk by candidate C

$$\text{Department D}$$

Hire	Don't Hire
4, 4	0, 0

Fig. 6.12 Bimatrix game with candidate C's type: qualified

$$\text{Department D}$$

Hire	Don't Hire
6, −2	0, 3

Fig. 6.13 Bimatrix game with candidate C's type: under-qualified

The outcome can move away from babbling equilibrium if candidate C signals costly effort before communicating her type to department D. Education is used as a signal of effort, which enhances the credibility of communication as shown in Fig. 6.14.

Cheap talk is extensively used by defenders in the entry-deterrence game (refer Example 5.1). The incumbents can use cheap talk to scare away potential entrants planning to enter the market. This can be seen in cases where an incumbent announces its willingness to enter a price war, in case the entrant decides to move into the market. There can be no way for the entrant to verify whether the incumbent will fight or acquiesce. Cheap talk is precisely a *cheap* tactic the incumbents use to deter potential entrants. Tingley and Walter (2011) perform an experimental study and find evidence that costless verbal communication does deter entrants in the early periods of play.

6.3 Cheap Talk in Operations Management

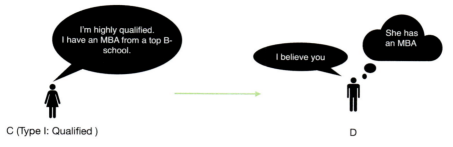

Fig. 6.14 Candidate C's MBA as a signal of effort

Another example of cheap talk is an airline may attract customers by claiming to have a high punctuality rate but may conveniently hide the fact that the duration stated on the ticket is larger than the actual duration.[9]

An interesting application of cheap talk is how customer service calls are handled. Ideally, a firm (or service provider) wants to match the number of customers with the number of service lines available. If the number of customers exceeds the service capacity, the firm risks losing goodwill as the customers have to wait before being served. To reduce the number of customers in the queue, the firms tend to make some delay announcements, hoping that some of the waiting customers may leave, reducing the burden on the service. Allon et al. (2011) study the dynamics of the influence of delay announcements on customer behavior. It becomes important for firms to understand the ways in which they can use delay announcements as "cheap talk" and elicit favorable behavior from the customers.

Consider a situation when the service provider follows a $M/M/1$ queue with the customers' arrival follows a Poisson process with parameter λ. The customer service times can be assumed to follow an exponential distribution with a mean of $1/\mu$, where μ is the service capacity—how many customers a server can serve per unit of time. The customers are *symmetric*, where each customer gets a utility of R on getting the service and incurs a cost of c per unit of time spent in the system (sojourn time). A customer can decide on joining the queue or balking, represented by the variable y. The utility function of a customer is given in Eq. 6.16.

$$U(y, w) = \begin{cases} R - cw & \text{if } y = \text{"join"} \\ 0 & \text{if } y = \text{"balk"} \end{cases} \quad (6.16)$$

where w denotes the customer's sojourn time in the system.

The service provider makes a gain of v for servicing a customer and incurs a cost of $h(w)$ for the customer's sojourn time w. The service provider has complete information about the system when a customer enters. On the arrival of a customer, the service makes a delay announcement, which the customer interprets as information

---

[9] *Airlines inflate flight durations: Flights "on time", yet passengers are delayed.* Indian Express, January 17, 2018.

Fig. 6.15 Joining and waiting

Fig. 6.16 Balking

θ. A customer tries to maximize his expected payoff and joins the queue only if $R \geq c\mathbb{E}(w|\theta)$. The delay announcement and its impact on the system depends both on the dynamics of the queuing system and also on the behavior of the customers. A message may tempt a customer to join the queue and wait, even if the queue is long (Fig. 6.15), or may nudge a customer towards leaving (or *balking*) even if the queue length is small (Fig. 6.16). The goals of a customer and the service provider are not completely misaligned—they want to minimize waiting time to maximize their payoffs.

The service provider can decide on a set of messages for announcement based on the queue length $\mathcal{Z}$. Let $\mathcal{M} = \{m_0, m_1, m_2, \ldots\}$ be the set of announcement messages the service provider can select using the signaling rule $g : \mathcal{Z} \mapsto \mathcal{M}$. The customers are indistinguishable and symmetric, and use a strategy $y : \mathcal{M} \mapsto \{0, 1\}$. The decision to "join" is denoted by $y(m) = 1$ and $y(m) = 0$ as "balk". Hence, the birth rate of the queue is given by $\lambda y(g(q))$, while the death rate is given by $\mu I_{q>0}$, where $I\{\cdot\}$ is the indicator function. The steady-state probability of having q customers in the system is given by $p_q(y, g)$.

The game between the service provider and the customers is solved using *Markov-Perfect Bayesian Nash Equilibrium* (MPBNE) as the solution concept. This equilibrium is an extension of Markov-Perfect Equilibrium (MPE), a refinement of SPNE. In MPE, the equilibrium rules depend only upon the current values of the state variables. The term perfect, as before, implies conformity with the procedure of backward induction and denotes optimality in all future possible states. The signaling rule $g(\cdot)$ and the action rule $y(\cdot)$ constitute MPBNE if the following conditions are satisfied (refer Allon et al., 2011):

6.4 Extensive-Form Inventory Games

i For each $\mathcal{M}$,

$$y(m) = \begin{cases} 1 & \frac{\sum_{\{q:g(q)=m\}}[R-c((q+1)/\mu)]p_q(y,g)}{\sum_{\{q:g(q)=m\}} p_q(y,g)} \geq 0 \\ 0 & \text{otherwise} \end{cases}$$

ii With $f(j) = v - \mathbb{E}[h(W(j+1))]$, where $W(j+1)$ is the time spent by a customer who joins the system with j customers, there exist constants $J_0, J_1, \ldots, J_q, \ldots$, and γ that satisfy the following set of equations:

$$J_0 = \max_{m \in \mathcal{M}} \left\{ \frac{f(0)y(m) - \gamma}{\lambda} + J_0(1 - y(m)) + J_1 y(m) \right\}$$

$$= \max_{m \in \mathcal{M}} \left\{ \frac{f(q)y(m) - \gamma}{\lambda + \mu} + \frac{\mu}{\lambda + \mu} J_{q-1} + \frac{\lambda}{\lambda + \mu} (J_q(1 - y(m)) + J_{q+1} y(m)) \right\}$$

The first condition states that, for a given signaling rule and when the signal is m, a customer adheres to the recommendation of the action rule if the expected conditional utility is positive. Likewise, given action rule y, the service provider adheres to its signaling rule g if its steady-state profits are maximized. For this, the signaling rule g is a solution of the Markov decision process in the second condition. The constant γ represents the long-term average profit made by the firm under optimal policy, and constants $J_0, J_1, \ldots, J_q, \ldots$ represent the relative cost for the states $0, 1, \ldots, q, \ldots$, respectively. The results show that even when the information provided to the customers is non-verifiable, the payoffs of the service provider and the customers improve by cheap talk.

6.4 Extensive-Form Inventory Games[10]

We discussed inventory games using normal-form representation in Sect. 4.1, where we focused on newsvendor games with simultaneous inventory decisions of multiple newsvendors. However, numerous real-world scenarios involve situations where one newsvendor moves before the other—one decides the order quantity before the other newsvendor. The first-mover newsvendor is P1, while the second mover is P2. This newsvendor setting is similar to the *von Stackelberg duopoly* (refer Example 5.3). As illustrated in Fig. 6.17, the setting can be modeled as extensive-form representation with subgame perfect Nash equilibrium (SPNE) as the reasonable solution concept. Backward induction can be used to compute subgame perfect Nash equilibrium (SPNE) (Wu & Parlar, 2011).

The expected profit functions devised for players in Sect. 4.1 carry over to this setting, and only the solution methodology differs. Instead of simultaneously solving

---

[10] In this section, the OM actions are on the *asset side* in the VCAP framework.

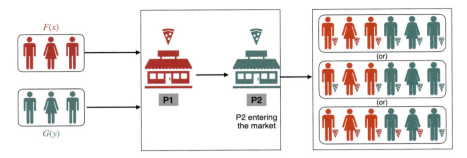

Fig. 6.17 Extensive-form newsvendor game

the problem, we can use backward induction. P2, observing the choice of P1, decides his best response n^*. Conjecturing the best response of P2, P1 then decides her optimal order quantity m^*. The best response function $n_s^*(m)$ of P2 is obtained by maximizing $J_2(m, n)$ (Eq. 6.17).

$$n_s^*(m) = \arg\max J_2(m, n) = \left\{ n : \frac{\partial J_2(m, n)}{\partial n} = 0 \right\} \quad (6.17)$$

For P1, m^* is the solution of the optimization problem (Eq. 6.18).

$$\max J_1(m, n) \text{ subject to } n = n_s^*(m) \quad (6.18)$$

Figure 6.18 shows the projection of $J_1(m, n)$ on (m, n) plane. Also, note that Fig. 6.18 shows the reaction curves (or best response function) $n_s^*(m)$ of P2 (refer

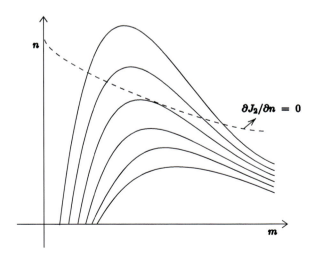

Fig. 6.18 Reaction curve (Adapted from Wu and Parlar 2011)

Eq. 6.17). The optimal m_s^* can be found by choosing the maximum valued contour that is tangent to the best response function $n_s^*(m)$ of P2.

Using the same data as in Sect. 4.1, $(m_s^*, n_s^*) = (28.38, 18.60)$ and the expected profits are 84.35 and 33.94 for players P1 and P2, respectively. A quick comparison between the profits reveals that $J_1(m_s^*, n_s^*) \geq J_1(m^*, n^*)$ and $J_2(m_s^*, n_s^*) \leq J_2(m^*, n^*)$, which implies P1 enjoys the first-mover advantage when the newsvendors move sequentially.

Problems

Problem 6.1 Consider the capacity decision problem discussed in Sect. 6.1, in which the developer outsources the manufacturing of doses. How the capacity decisions change if the developer is capable of manufacturing the vaccines on its own? In such a vertically integrated company I, the R&D division develops the vaccine, and the manufacturing division decides the at-risk quantity to be produced. Further, if the company's vaccine candidate fails to obtain regulatory approval, the manufacturing division can produce vaccine doses for another developer. Let γ_I be the capacity building coefficient and c_I be the unit production cost. Consider other parameters to be unchanged. ◀

Problem 6.2 In Sect. 6.1, consider the role of a government in providing funding to receive the target number of doses. The government can provide funding to either the developer or the manufacturer. If funding is provided to the developer, the probability of success increases for the developer. Alternatively, the government can provide funding for sharing the costs of at-risk manufacturing, with a lower limit to the at-risk capacity being set up $\underline{y}$. Let $\underline{y} = \sqrt{C_M/q^2\gamma}$, where q is the required capacity, C_M is the government's funding support to the manufacturer, and γ is capacity building coefficient. The government works under a fixed budget and must decide upon the division of this budget towards providing support for either the developer (C_D) or the manufacturer (C_M). Formulate the government's problem of maximizing the at-risk capacity. ◀

Problem 6.3 In Sect. 6.6, we discussed the changing profits for each node of the supply chain when the position of the leader is changed. Formulate the profit function of each node when:

  i retailer R is the leader,
 ii manufacturer M is the leader.

Compare the profits of different nodes. Is there an optimal position for the leader? An optimal position of the leader refers to a position in which the profits of all the nodes in the supply chain are maximized. ◀

References

Allon, G., Bassamboo, A., & Gurvich, I. (2011). "we will be right with you": Managing customer expectations with vague promises and cheap talk. *Operations research, 59*(6), 1382–1394.

Aumann, R. J., & Hart, S. (2003). Long cheap talk. *Econometrica, 71*(6), 1619–1660.

Kremer, M., & Glennerster, R. (2004). *Strong medicine: Creating incentives for pharmaceutical research on neglected diseases*. Princeton University Press.

Kremer, M., Levin, J., & Snyder, C. M. (2020). Advance market commitments: insights from theory and experience. In *AEA Papers and Proceedings* (vol. 110, pp. 269–273). American Economic Association 2014 Broadway, Suite 305, Nashville, TN 37203.

Majumder, P., & Srinivasan, A. (2006). Leader location, cooperation, and coordination in serial supply chains. *Production and Operations Management, 15*(1), 22–39.

Martin, P., Gupta, D., & Natarajan, K. V. (2020). Vaccine procurement contracts for developing countries. *Production and Operations Management, 29*(11), 2601–2620.

Song, J.-S., Van Houtum, G.-J., & Van Mieghem, J. A. (2020). Capacity and inventory management: Review, trends, and projections. *Manufacturing & Service Operations Management, 22*(1), 36–46.

Spence, M. (1978). Job market signaling. In *Uncertainty in economics* (pp. 281–306). Elsevier.

Spengler, J. J. (1950). Vertical integration and antitrust policy. *Journal of Political Economy, 58*(4), 347–352.

Sun, H., Toyasaki, F., & Falagara Sigala, I. (2023). Incentivizing at-risk production capacity building for covid-19 vaccines. *Production and Operations Management, 32*(5), 1550–1566.

Tingley, D. H., & Walter, B. F. (2011). Can cheap talk deter? an experimental analysis. *Journal of Conflict Resolution, 55*(6), 996–1020.

Van Mieghem, J. A. (2003). Commissioned paper: Capacity management, investment, and hedging: Review and recent developments. *Manufacturing & Service Operations Management, 5*(4), 269–302.

Wu, H., & Parlar, M. (2011). Games with incomplete information: A simplified exposition with inventory management applications. *International Journal of Production Economics, 133*(2), 562–577.

Chapter 7
Games in Characteristic Form

We introduced characteristic-form games $\langle N, v \rangle$ in Sect. 2.2.3, which is a representation of cooperative games. In these games, a characteristic function $v : 2^n \mapsto \mathbb{R}$ assigns a payoff to each coalition. $v(\emptyset) = 0$. n is the cardinality of the finite set N of players. For coalition $S \subseteq N$, $v(S)$ is the maximum payoff that coalition S can guarantee independent of the players in $N \setminus S$. In this chapter, we study reasonable solution concepts for characteristic-form games. We restrict our focus on superadditive games in which $v(S \cup T) \geq v(S) + v(T)$ for each disjoint sets $S, T \subset N$. The superadditivity condition can be justified when coalitions S and T complement each other. For superadditive games, the grand coalition of all players in N achieves the maximum payoff.

Solution concepts for superadditive games prescribe the division of payoff among the agents in the grand coalition N and can be broadly classified into

i. *Stability based*—how to divide the payoff to ensure the stability of the grand coalition? The *core* is an example of such a class of solution concepts.
ii. *Fairness based*—how to divide the payoff to achieve fairness? The *Shapley value* is an example of this class of solution concepts.

7.1 Examples

We discuss some examples of characteristic-form games for better understanding.

Example 7.1 (*Voting Game*) Suppose in a parliament with a strength of 100, political parties X, Y, and Z have 45, 35, and 25 members in the parliament, respectively. A bill, which needs a minimum of two-thirds of votes to pass, gives a value of 1 to the winning coalition and 0 otherwise. Hence, $v(\{X\}) = v(\{Y\}) = v(\{Z\}) = v(\{Y, Z\}) = 0$, $v(\{X, Y\}) = v(\{X, Z\}) = v(\{X, Y, Z\}) = 1$. This is an example of *simple game* for which $v(S) \in \{0, 1\}$ for each $S \subseteq N$. X is a *veto player*, as it belongs to all the winning coalitions. ◁

© The Author(s), under exclusive license to Springer Nature Singapore Pte Ltd. 2024   125
R. K. Amit, *Game Theory with Applications in Operations Management*, Springer Texts in Business and Economics, https://doi.org/10.1007/978-981-99-4833-8_7

Example 7.2 (*Recycling Game*) In recent years, there has been an emphasis on extended producer responsibility (EPR), under which producers are responsible for the entire lifecycle of the products, including end-of-life products. Many countries have passed legislation for EPR for end-of-life vehicles (ELV) management (Mohan & Amit, 2020).

Consider a setting where three automobile manufacturers, A, B, and C, are mandated by the EPR law to recycle their products. Each manufacturer can decide to build an independent recycling plant with a stand-alone cost of $c(\{i\})$. They can form bilateral[1] or trilateral alliances (or coalitions) S for cooperative recycling with cost $c(S)$. For example, S can be coalition $\{A, B\}$, or $\{A, C\}$, or $\{B, C\}$, or the grand coalition $\{A, B, C\}$. We preclude the possibility of overlapping coalitions.[2] The setting can be represented as a characteristic-form game called *recycling game*. The characteristic function is the cost function defined for each $S \subseteq N$ as shown below (all numbers are in Indian rupee billions):

$$c(\emptyset) = 0$$
$$c(\{A\}) = 4 \quad c(\{B\}) = 2 \quad c(\{C\}) = 3$$
$$c(\{A, B\}) = 5 \quad c(\{A, C\}) = 6 \quad c(\{B, C\}) = 4$$
$$c(\{A, B, C\}) = 7$$

Based on the cost information, the given recycling game is *subadditive*, that is, for each disjoint sets $S, T \subset N$, $c(S) + c(T) \geq c(S \cup T)$ (compare the subadditivity condition with the superadditivity condition). As there are scale effects—cost decreases with a larger coalition, and hence the grand coalition has the least cumulative cost—a relevant question is how to share the cost of the joint recycling plant among A, B, and C such that the trilateral alliance remains stable. $\triangleleft$

Example 7.3 (*Trade Game*) Consider a setting where a seller A owns a good that is worthless to her. She identifies two potential buyers, B and C, who value the good at ₹70 million and ₹100 million, respectively. If the trade happens between A and B, the value is ₹70 million, with some payment from B to A. The setting can be represented as a characteristic-form game called *trade game*, with the following characteristic function:

$$v(\emptyset) = 0$$
$$v(\{A\}) = 0 \quad v(\{B\}) = 0 \quad v(\{C\}) = 0$$
$$v(\{A, B\}) = 70 \quad v(\{A, C\}) = 100 \quad v(\{B, C\}) = 0$$
$$v(\{A, B, C\}) = 100$$

The efficient outcome (social-welfare maximizing) is when the good is sold to the buyer C. It would be interesting to compute the payment from C to A, which makes the grand coalition stable—there is no incentive for A to ignore the valuation of B when deciding to sell the good to C. It can be verified that the minimum payment from

---

[1] Suzuki, Toyota allies at Indian ELV recycling facility, *Recycling Today*, November 24, 2021.

[2] We recommend (Mahdiraji et al., 2021) for a review of cooperative games with overlapping coalitions.

7.1 Examples

C to A is ₹70 million, which ensures the stability of the grand coalition. Compare the answer with the outcome of the second-price sealed-bid auction (refer to Example 3.10) when the valuations are private information. ◁

Example 7.4 (*Linear Production Game*) Owen (1975) introduce the linear production game. In the linear production game with N players, each player i owns a bundle of $(b_{i1}, \ldots, b_{iq})$ of q commodities where b_{i1} is the number of units of commodity C_1. The commodities have no value on their own; however, they can be used to produce goods $G_1, \ldots, G_m$ using a linear production technology. Each unit of good G_l has a market price p_l and uses a_{1l} of commodity $C_1, \ldots, a_{ql}$ of commodity C_q, as shown in Fig. 7.1.

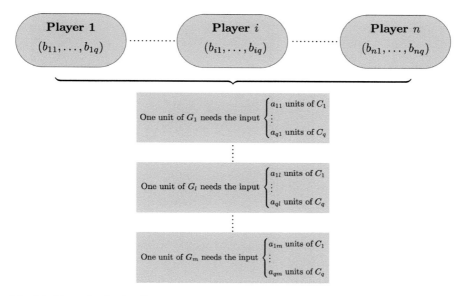

Fig. 7.1 Linear Production Game

The players can form coalitions and pool their bundles to produce different goods. The characteristic function for coalition $S \subseteq N$ is

$$v(S) = \max \sum_{l=1}^{m} p_l x_l$$

$$\text{s.t.} \quad \sum_{l=1}^{m} a_{kl} x_l \leq \sum_{i \in S} b_{ik} \quad \forall k = 1, \ldots, q$$

$$x_l \geq 0 \quad \forall l = 1, \ldots, m$$

where x_l is the amount of good G_l produced by the pooling of bundles. The linear production game is superadditive. A relevant question is how to divide the worth of the grand coalition $v(N)$ to make the grand coalition stable. ◁

7.2 Some Definitions for Characteristic-Form Games

We defined characteristic-form games $\langle N, v \rangle$ and superadditive games in Sect. 2.2.3. We also mentioned that, for superadditive characteristic-form games, the grand coalition achieves the maximum payoff. In this section, we introduce further concepts that are relevant for discussing solution concepts for characteristic-form games.

Definition 7.1 (*Inessential and Essential Games*) A characteristic-form game $\langle N, v \rangle$ is *inessential* if $\sum v(\{i\}) = v(N)$, and *essential* if $\sum v(\{i\}) < v(N)$. ◀

Definition 7.2 (*Constant-sum Games*) A characteristic-form game $\langle N, v \rangle$ is *constant-sum* if $\forall S \subset N, v(S) + v(N \setminus S) = v(N)$. ◀

An inessential game is always a constant-sum game, but the converse is not true. In other words, Inessential Game $\subset$ Constant-sum Game.

Definition 7.3 (*Convex Games*) A characteristic-form game $\langle N, v \rangle$ is *convex* if $\forall S \subset N, v(S \cup \{i\}) - v(S)$ is increasing in $|S|^3$ for each i. ◀

$v(S \cup \{i\}) - v(S)$ is the marginal contribution of player i to coalition S. In convex games, as the marginal contribution of player i increases with the size of the coalition, player i's incentive to join a coalition increases with the size of the coalition. Convex games capture the concept of *complementarity* that has wide practical applications.

A convex characteristic-form game $\langle N, v \rangle$ with $v(\emptyset) = 0$ is superadditive. We assumed $v(\emptyset) = 0$ while defining characteristic-form games in Definition 2.2.3; hence, in this book, each convex game is superadditive, but the converse is not true, as illustrated in the following example.

Example 7.5 Consider characteristic-form game $\langle N, v \rangle$ with $N = \{1, 2, 3\}$ with the characteristic function defined as

$$
v(S) = \begin{cases}
0 & |S| = 0 \\
1 & |S| = 1 \\
3 & |S| = 2 \\
4 & |S| = 3
\end{cases}
$$

It can be verified that the game is superadditive; however, it is not a convex game. For $i = 1$ and $S = \{2\}$, $v(S \cup \{i\}) - v(S) = 3 - 1 = 2$. For $i = 1$ and $S = \{2, 3\}$, $v(S \cup \{i\}) - v(S) = 4 - 3 = 1$. This means $v(S \cup \{i\}) - v(S)$ is not an increasing function in $|S|$; hence the game is not convex. ◁

Solution concepts for superadditive characteristic-form games prescribe the division of payoff among the agents in the grand coalition N. We define payoff vector

---

^{3} Cardinality or size of S is represented as $|S|$.

7.2 Some Definitions for Characteristic-Form Games

$\mathbf{x} = (x_1, \ldots, x_n)$ with x_i as the share of player i in the grand coalition's payoff $v(N)$. Some definitions for payoff vectors are discussed next.

Definition 7.4 (*Feasible Payoff*) A payoff vector $\mathbf{x} = (x_1, \ldots, x_n)$ is said to be *feasible* if $\sum x_i \leq v(N)$. ◂

Definition 7.5 (*Efficient Payoff*) A payoff vector $\mathbf{x} = (x_1, \ldots, x_n)$ is said to be *efficient* if $\sum x_i = v(N)$. ◂

Definition 7.6 (*Individually Rational Payoff*) A payoff vector $\mathbf{x} = (x_1, \ldots, x_n)$ is said to be *individually rational* if $x_i \geq v(\{i\})$. ◂

Definition 7.7 (*Imputation*) An *imputation* is a payoff vector that is efficient and individually rational. ◂

Example 7.6 Consider characteristic-form game $\langle N, v \rangle$ with $N = \{1, 2, 3\}$ with the characteristic function defined as

$$v(\{1\}) = 1;\ v(\{2\}) = 0;\ v(\{3\}) = 1$$
$$v(\{1, 2\}) = 4;\ v(\{1, 3\}) = 3;\ v(\{2, 3\}) = 5$$
$$v(\{1, 2, 3\}) = 8$$

The set of imputations for the game is shown in Fig. 7.2 as the plane (shaded triangle) in which $x_1 + x_2 + x_3 = 8$.

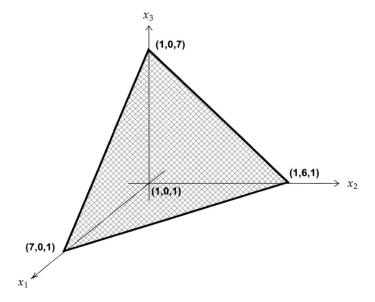

Fig. 7.2 Set of Imputations ◂

130   7 Games in Characteristic Form

7.3 Solution Concepts

In this section, we discuss some important stability-based and fairness-based solution concepts for superadditive characteristic-form games.

7.3.1 The Core

The core is one of the fundamental stability-based solution concepts. For superadditive characteristic-form games, it prescribes the division of payoff among the agents in the grand coalition such that no sub-coalition *blocks* the division—no sub-coalition has the incentive to come out of the grand coalition. This means that the core ensures the stability of the grand coalition (or the division agreement).

Definition 7.8 (*Core*) The core is a set of imputations $\mathbf{x} = (x_1, \ldots, x_n)$ for characteristic-form game $\langle N, v \rangle$ such that $\forall S \subseteq N, \sum_{i \in S} x_i \geq v(S)$.   ◄

For noncooperative games, Nash equilibrium is the most important solution concept. We know that in Nash equilibrium, unilateral deviations are suboptimal. The core is an analogous solution concept for cooperative games in which coalitional deviations are suboptimal (no sub-coalition has the incentive to deviate from an imputation in the core). Like in noncooperative game theory, a game can have multiple Nash equilibria, and the core in a characteristic-form game can have multiple imputations. But, unlike the guaranteed existence of Nash equilibrium, at least in mixed strategies, the core in a characteristic-form game can be empty. This motivates research in other solution concepts for cooperative games that prescribe a unique payoff vector like the *nucleolus*.

The following examples improve understanding of the core.

Example 7.7 Consider characteristic-form game of Example 7.6, $\langle N, v \rangle$ with $N = \{1, 2, 3\}$ with the characteristic function defined as

$$v(\{1\}) = 1; \ v(\{2\}) = 0; \ v(\{3\}) = 1$$
$$v(\{1, 2\}) = 4; \ v(\{1, 3\}) = 3; \ v(\{2, 3\}) = 5$$
$$v(\{1, 2, 3\}) = 8$$

From Definition 7.8, the core is the intersection of linear inequalities $\sum_{i \in S} x_i \geq v(S)$. For this game, the linear inequalities that define the core are

$$x_1 \geq 1; \ x_2 \geq 0; \ x_3 \geq 1$$
$$x_1 + x_2 \geq 4; \ x_1 + x_3 \geq 3; \ x_2 + x_3 \geq 5$$
$$x_1 + x_2 + x_3 = 8$$

Combining the set of imputations in Fig. 7.2 and the linear inequalities, the core imputations are shown in yellow color in Fig. 7.3.

7.3 Solution Concepts

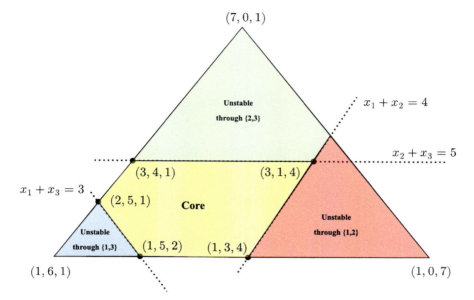

Fig. 7.3 The core

The set of imputations that can be blocked by the coalition $\{1, 2\}$, coalition $\{1, 3\}$, and $\{2, 3\}$ are shown in red, blue, and green color, respectively. ◁

Example 7.8 (*Core of the Trade Game*) We discussed the trade game in Example 7.3 and mentioned that the minimum payment from C to A is ₹70 million to ensure the stability of the grand coalition. We show that this condition is part of the core of the trade game. The characteristic function is

$$v(\emptyset) = 0$$
$$v(\{A\}) = 0 \quad v(\{B\}) = 0 \quad v(\{C\}) = 0$$
$$v(\{A, B\}) = 70 \quad v(\{A, C\}) = 100 \quad v(\{B, C\}) = 0$$
$$v(\{A, B, C\}) = 100$$

A payoff vector $\mathbf{x} = (x_A, x_B, x_C)$ lies in the core of the game if the following conditions are satisfied:

$$x_A \geq 0; \quad x_B \geq 0; \quad x_C \geq 0$$
$$x_A + x_B \geq 70; \quad x_B + x_C \geq 0; \quad x_A + x_C \geq 100$$
$$x_A + x_B + x_C = 100$$

The conditions lead to $x_A \geq 70$, $x_B = 0$, $x_C \geq 0$, and $x_A + x_C = 100$. This means, in the core imputations, any payment above ₹70 million from C to A ensures the stability of the grand coalition. ◁

Example 7.9 Consider characteristic-form game of Example 7.5, $\langle N, v \rangle$ with $N = \{1, 2, 3\}$ and the characteristic function defined as

$$v(S) = \begin{cases} 0 & |S| = 0 \\ 1 & |S| = 1 \\ 3 & |S| = 2 \\ 4 & |S| = 3 \end{cases}$$

A payoff vector $\mathbf{x} = (x_1, x_2, x_3)$ lies in the core of the game if the following conditions are satisfied:

$$x_1 \geq 1; \quad x_2 \geq 1; \quad x_3 \geq 1$$
$$x_1 + x_2 \geq 3; \quad x_2 + x_3 \geq 3; \quad x_1 + x_3 \geq 3$$
$$x_1 + x_2 + x_3 = 4$$

Adding the inequalities in the second row gives $x_1 + x_2 + x_3 \geq 4.5$, which violates the condition $x_1 + x_2 + x_3 = 4$. There does not exist any imputation that satisfies the conditions of the core—*the core is empty.* ◁

We know that the game in Example 7.5 is not convex. Is there a relationship between convex games and the nonemptiness of the core? In the next section, we discuss the result that *the core of a convex game is nonempty.*

Balancedness is another property that can be used to characterize the nonemptiness of the core. Let $\mathcal{S}$ be the set of possible coalitions $S \subseteq N$. The cardinality of set $\mathcal{S}$ is 2^n.

Definition 7.9 (*Balanced Collection*) Set $\mathcal{S}$ is a *balanced collection* if there exists a set of nonnegative numbers κ_S for each $S \in \mathcal{S}$ such that

$$\sum_{\substack{S \in \mathcal{S} \\ i \in S}} \kappa_S = 1, \quad \forall i \in N$$

κ_S are called *balancing weights* of collection $\mathcal{S}$. ◀

κ_S can be interpreted as a probability assigned by player i that coalition $S \in \mathcal{S}$ will form, when he is a member of S.

Definition 7.10 (*Balanced Game*) A characteristic-form game $\langle N, v \rangle$ is a *balanced game* if for each set of balancing weights κ_S of balanced collection $\mathcal{S}$, the following condition is satisfied:

$$v(N) \geq \sum_{S \in \mathcal{S}} \kappa_S \cdot v(S)$$ ◀

Theorem 7.1 (Bondareva–Shapley Theorem Bondareva, 1963; Shapley, 1967) *A characteristic-form game $\langle N, v \rangle$ has a nonempty core if and if only it is balanced.* ◀

7.3 Solution Concepts 133

Proof To compute the core, consider the following linear programming formulation

$$\min_{\substack{\mathbf{x} \\ \sum x_i \leq v(N)}} \quad \sum_{i \in N} x_i$$

$$\text{s.t.} \quad \sum_{i \in S} x_i \geq v(S), \quad \forall S \subseteq N$$

where $\mathbf{x}$ is a feasible payoff vector (Definition 7.4). The core is nonempty if and only if the solution of the above formulation is also an efficient payoff vector ($\sum_{i \in N} x_i = v(N)$).

Consider the dual of the above linear programming formulation

$$\max_{\kappa_S} \quad \sum_{S \in \mathcal{S}} \kappa_S \cdot v(S)$$

$$\text{s.t.} \quad \sum_{\substack{S \in \mathcal{S} \\ i \in S}} \kappa_S = 1, \quad \forall i \in N$$

$$\kappa_S \geq 0, \qquad \forall S \in \mathcal{S}$$

The first constraint in the dual formulation ensures $\mathcal{S}$ is a balanced collection with balancing weights κ_S. Using weak duality lemma (Luenberger & Ye, 2008), the optimal value of the dual problem is bounded above by the optimal value of the primal problem, which means $v(N) \geq \sum_{S \in \mathcal{S}} \kappa_S \cdot v(S)$. Hence, the game is balanced when the core is nonempty (the optimal value of the primal problem when the core is nonempty is $v(N)$).

If the game is balanced, then $v(N) \geq \sum_{S \in \mathcal{S}} \kappa_S \cdot v(S)$, which means that the optimal solution of the dual problem is bounded above by $v(N)$. If $v(N)$ is the optimal value of the primal problem, then the core is nonempty. This completes the proof. $\qquad \square$

7.3.2 Shapley Value

We discussed the core as one of the solution concepts that prescribes the set of imputations to ensure the stability of the grand coalition. In this section, we discuss a fairness-based solution concept called the *Shapley value*.

We discuss the desiderata before defining the Shapley value. For a characteristic-form game $\langle N, v \rangle$, let π be a *permutation* of $N = \{1, \ldots, n\}$, and Π be the set of all such $n!$ permutations of N.

For permutation $\pi \in \Pi$, $S(\pi, i)$ is the coalition of the first i players in permutation π, and $S(\pi, 0) = \emptyset$

$$S(\pi, i) = \{\pi(1), \ldots, \pi(i)\} \ \text{ for } \ i = 1, \ldots, n$$

Shapley (1971) introduces the *greedy algorithm* for cooperative games. For permutation π, the greedy algorithm provides the payoff vector $\mathbf{x}(\pi)$, in which the player at position $\pi(i)$ gets a payoff $x_{\pi(i)}$ that is equal to the marginal contribution of the player in permutation π. Mathematically,

$$x_{\pi(i)} = v(S(\pi, i)) - v(S(\pi, i-1)) \text{ for } i = 1, \ldots, n$$

It can be verified that $\mathbf{x}(\pi)$ is an efficient payoff vector.

Example 7.10 Consider characteristic-form game of Example 7.6, $\langle N, v \rangle$ with $N = \{1, 2, 3\}$ with the characteristic function defined as

$$v(\{1\}) = 1; v(\{2\}) = 0; v(\{3\}) = 1$$
$$v(\{1, 2\}) = 4; v(\{1, 3\}) = 3; v(\{2, 3\}) = 5$$
$$v(\{1, 2, 3\}) = 8$$

The total number of possible permutations is $3! = 6$ as shown in Fig. 7.4, with π as a specific permutation of players.

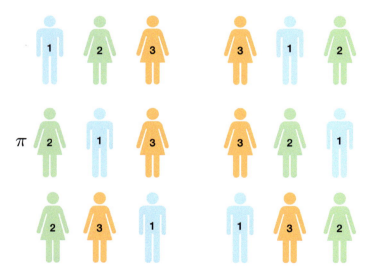

Fig. 7.4 Permutations of Players

For the permutation π, $S(\pi, 1) = \{2\}$ as player 2 is at the first position in the permutation. Similarly, $S(\pi, 2) = \{2, 1\}$ and $S(\pi, 3) = \{2, 1, 3\}$.

Using the greedy algorithm, for permutation π, $x_{\pi(1)}$ is the payoff of player 2.

7.3 Solution Concepts

$$x_{\pi(1)} = v(S(\pi, 1)) - v(S(\pi, 0))$$
$$= v(\{2\}) - v(\emptyset)$$
$$= 0$$

We can check that $x_{\pi(2)}$ (payoff of player 1)= 4 and $x_{\pi(3)}$ (payoff of player 3)= 4, and $\mathbf{x}(\pi) = (0, 4, 4)$. ◁

With this background, we now discuss the Shapley value. Shapley (1953) proposes the *Shapley value* $\Phi(v) = (\phi_1(v), ..., \phi_i(v), ..., \phi_n(v))$, a fairness-based solution concept for cooperative games. The Shapley value $\Phi(v)$ is the unique payoff vector that satisfies the following axioms, which characterize fairness:

Axiom I: **Efficiency** $\sum_{i \in N} \phi_i(v) = v(N)$.

Axiom II: **Symmetry** $\phi_i(v)$ is not affected by permutations of players.

Axiom III: **Dummy Player** If $v(S) = v(S \cup \{i\}) \, \forall S \subset N \setminus \{i\}$, then i is a dummy player and $\phi_i(v) = 0$.

Axiom IV: **Additivity** If $\langle N, v_1 \rangle$ and $\langle N, v_2 \rangle$ are two characteristic-form games, and $\langle N, v_1 + v_2 \rangle$ is the aggregated characteristic-form game with $(v_1 + v_2)(S) = v_1(S) + v_2(S) \, \forall S \subset N$, then $\phi_i(v_1 + v_2) = \phi_i(v_1) + \phi_i(v_2)$.

Theorem 7.2 (Shapley, 1953) *The Shapley value $\Phi(v)$ is the unique payoff vector for characteristic-form game $\langle N, v \rangle$ that satisfies Axioms I–IV. The Shapley value for player i is*

$$\phi_i(v) = \sum_{\substack{i \in S \\ S \subseteq N}} \frac{1}{n} \binom{n-1}{s-1}^{-1} [v(S) - v(S - \{i\})] \qquad ◀$$

The Shapley value is based on the evaluation of *prospects* of playing a game (Shapley, 1953, p. 307) notes.

> At the foundation of the theory of games is the assumption that the players of a game can evaluate, in their utility scales, every "prospect" that might arise as a result of play. In attempting to apply the theory to any field, one would normally expect to be permitted to include, in the class of "prospects", the prospect of having to play a game. The possibility of evaluating games is therefore of critical importance.

Based on this principle, the Shapley value $\phi_i(v)$ is the *expected marginal contribution* of player i when the player i has the belief that the coalition she joins is *equally likely* to be any size $s - 1$ ($0 \leq s - 1 \leq n - 1$) and the coalitions of size s are *equally likely* (Harsanyi, 1977). We recommend (Amit & Ramachandran, 2013) for the proof of Theorem 7.2 and related discussion.

Example 7.11 Consider characteristic-form game of Example 7.6, $\langle N, v \rangle$ with $N = \{1, 2, 3\}$ with the characteristic function defined as

$$v(\{1\}) = 1; v(\{2\}) = 0; v(\{3\}) = 1$$
$$v(\{1, 2\}) = 4; v(\{1, 3\}) = 3; v(\{2, 3\}) = 5$$
$$v(\{1, 2, 3\}) = 8$$

In this game, the number of possible coalitions is $2^3 = 8$. In half of them, player 1 is not a member, and the marginal contribution of player 1 to each such coalition is shown in Fig. 7.5, along with the probability of each coalition. The Shapley value is the expected marginal contribution.

$$\phi_1 = \frac{1}{3} \times 1 + \frac{1}{6} \times 4 + \frac{1}{6} \times 2 + \frac{1}{3} \times 3 = \frac{14}{6}$$

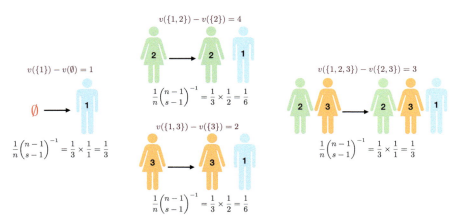

Fig. 7.5 Marginal contributions of player 1 to different coalitions and their probabilities

Similarly, we can compute $\phi_2 = \frac{17}{6}$ and $\phi_3 = \frac{17}{6}$. ◁

The Shapley value can also be computed using the greedy algorithm. ϕ_i is the average of the marginal contribution of player i in each permutation. The computations are shown in Fig. 7.6.

The Shapley value is a measure of the importance of a player in a game. In Example 7.11, players 2 and 3 are more important than player 1, as their Shapley value is higher. We recommend (Owen, 2001) for discussion on the indices of power, including the Shapley values as a power index.

The following theorem connects the core, the Shapley value, and the convex games.

7.3 Solution Concepts 137

Permutation ↓	Marginal Contribution of Player		
	Player 1	Player 2	Player 3
1 2 3	1	3	4
1 3 2	1	5	2
2 1 3	4	0	4
2 3 1	3	0	5
3 1 2	2	5	1
3 2 1	3	4	1
$\phi_i \longrightarrow$ Average Marginal Contribution	14/6	17/6	17/6

Fig. 7.6 Computing the Shapley value using the greedy algorithm

Theorem 7.3 (Shapley, 1971) *For **convex** characteristic-form game* $\langle N, v \rangle$,

a. *The payoff vector* $\mathbf{x}(\pi)$ *generated by the greedy algorithm for permutation* π *is in the core of the game.*
b. *The core is nonempty.*
c. *The Shapley value is in the core of the game.* ◀

Proof We discuss the proof idea using the modified version of game in Example 7.6. It can be checked that the characteristic-form game in Example 7.6 is not a convex game,[4] and the payoffs are modified to ensure the convexity of the game. The modified game is $\langle N, u \rangle$ with $N = \{1, 2, 3\}$ with the characteristic function defined as

$$u(\{1\}) = 1; u(\{2\}) = 0; u(\{3\}) = 1$$
$$u(\{1, 2\}) = 4; u(\{1, 3\}) = 3; u(\{2, 3\}) = 5$$
$$u(\{1, 2, 3\}) = 10$$

The marginal contribution of each player in each possible permutation using the greedy algorithm, along with the Shapley values, is shown in Fig. 7.7.

For permutation π', the greedy algorithm generates the payoff vector $\mathbf{x}(\pi') = (1, 2, 7)$. The payoff to player 2 is 7.

Let us consider sub-coalition $S' = \{1, 2\}$. In permutation π', $\sum_{i \in S'} x_i = x_1 + x_2 = 1 + 7 = 8$. $u(S')$ can be written as $u(\{1, 2\} - u(\{1\}) + u(\{1\}) - u(\emptyset) = 4$. $\sum_{i \in S'} x_i > u(S')$ because, in the convex game, player 2's marginal contribution is higher when it joins the larger coalition $\{1, 3\}$ in permutation π' compared to joining

---

[4] Convexity of characteristic-form games can be checked using CoopGame package in R.

Permutation	Marginal Contribution of Player		
	Player 1	Player 2	Player 3
1 2 3	1	3	6
1 3 2	1	7	2
2 1 3	4	0	6
2 3 1	5	0	5
3 1 2	2	7	1
3 2 1	5	4	1
ϕ_i Average Marginal Contribution	3	3.5	3.5

π' labels the permutation rows.

Fig. 7.7 Marginal contribution in each permutation using the greedy algorithm

$\{1\}$. This is true for any sub-coalition and any permutation in a convex game. Hence, $\sum_{i \in S} x_i \geq u(S)$ for each $S \subseteq N$ where x_i is an element of $\mathbf{x}(\pi')$. This proves part a. of the theorem. This also proves part b. of the theorem as at least one imputation exists in the core.

Furthermore, we know that using the greedy algorithm, the Shapley value ϕ_i is the average of the marginal contribution of player i in each permutation. For any sub-coalition $S' \subseteq N$

$$\sum_{i \in S'} \phi_i = \sum_{i \in S'} \sum_{\pi \in \Pi} \frac{1}{n!} x_{\pi(i)}$$

$$= \frac{1}{n!} \sum_{\pi \in \Pi} \sum_{i \in S'} x_{\pi(i)}$$

$$\geq \frac{1}{n!} \sum_{\pi \in \Pi} u(S') \quad \text{(from part a. of the theorem)}$$

$$\geq u(S')$$

This means that the Shapley value payoffs satisfy the core condition. This proves part c. of the theorem. $\qquad\square$

Shapley (1971) proves that the payoff vectors $\mathbf{x}(\pi)$, generated by the greedy algorithm for all $n!$ permutations, form the vertices of the core of a convex game. Also, the Shapley value is at the centroid of the core of a convex game. This is shown in Fig. 7.8 for the convex game discussed in Theorem 7.3. The core is shown in color, and it can be verified that the vertices are the 3! payoff vectors in Fig. 7.7.

7.3 Solution Concepts

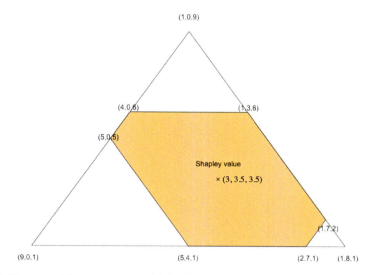

Fig. 7.8 The core of the convex game with the Shapley value as the centroid of the core

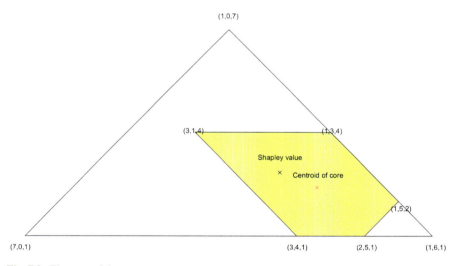

Fig. 7.9 The core of the nonconvex game

Ichiishi (1981) proves that if all $n!$ payoff vectors $\mathbf{x}(\pi)$ belong to the core, then the game is a convex game. We can illustrate this result using the game of Example 7.6. The core of the game is shown in color in Fig. 7.9. It can be observed that some $\mathbf{x}(\pi)$ (payoff vectors in Fig. 7.6) does not belong to the core, and hence the game is not convex. Also, the Shapley value lies in the core of the game, but does not coincide with the centroid of the core. As the core is nonempty and the Shapley value lies in the core for the game that is not convex, the conditions given in Theorem 7.3 are sufficiency conditions, but not necessary conditions, for the nonemptiness of the core. Bondareva–Shapley theorem (Theorem 7.1) provides the necessary and sufficient conditions for the nonemptiness of the core.

Problems

Problem 7.1 A customer wants to buy a phone and a sim card. There are three sellers, but Seller 1 has the phone, and Sellers 2 and 3 sell sim cards. A phone together with a sim card is worth ₹5,000 but is worthless otherwise. Also, a sim card without a phone is worthless.

  i Define a characteristic function for this game and verify that it is superadditive.
 ii Compute the core of the game. ◄

Problem 7.2 Consider game $\langle\{1, 2, 3, 4\}, v\rangle$ in which $v(1, 2, 3, 4) = 3$; $v(S) = 0$ if S includes at most one of the players in $\{1, 2, 3\}$, and $v(S) = 2$ otherwise. Compute the core and the Shapley value. ◄

Problem 7.3 Consider game $\langle\{1, 2, 3\}, v\rangle$ in which $v(1, 2, 3) = v(1, 2) = v(1, 3) = 1$ and $v(S) = 0$ otherwise. Is it a convex game? ◄

Problem 7.4 Consider the trade game in Example 7.3. Compute the Shapley value of the game. Does the Shapley value lie in the core of the game? ◄

References

Amit, R. K., & Ramachandran, P. (2013). Aspects of exchangeability in the Shapley value. *International Game Theory Review*.

Bondareva, O. N. (1963). Some applications of linear programming methods to the theory of cooperative games (In Russian). *Problemy Kybernetiki, 10*, 119–139.

Harsanyi, J. C. (1977). *Rational behavior and bargaining equilibrium in games and social situations*. Cambridge University Pr.

Ichiishi, T. (1981). Super-modularity: Applications to convex games and to the greedy algorithm for LP. *Journal of Economic Theory, 25*(2), 283–286.

Luenberger, D. G., & Ye, Y. (2008). *Linear and nonlinear programming, volume 116 of international series in operations research and management science*. Springer.

Mahdiraji, H. A., Razghandi, E., & Hatami-Marbini, A. (2021). Overlapping coalition formation in game theory: A state-of-the-art review. *Expert Systems with Applications, 174*, 114752.

Mohan, T. V., & Amit, R. K. (2020). Dismantlers' dilemma in end-of-life vehicle recycling markets: a system dynamics model. *Annals of Operations Research, 290*(1–2), 591–619.

Owen, G. (1975). On the core of linear production games. *Mathematical Programming, 9*(1), 358–370.

Owen, G. (2001). *Game theory*. Academic.

References

Shapley, L. S. (1953). A *value for n-person games. volume II of annals of mathematics studies*. Princeton University Press.

Shapley, L. S. (1967). On balanced sets and cores. *Naval Research Logistics Quarterly, 14*(4), 453–460.

Shapley, L. S. (1971). Cores of convex games. *International Journal of Game Theory, 1*(1), 11–26.

Chapter 8
Games in Characteristic Form: Applications in OM

In Chap. 7, we discussed games in characteristic form, and important solution concepts for that class of games. This chapter provides an exposition of numerous operations management (OM) applications[1] of games in characteristic form.

8.1 Inventory Centralization in Supply Chains[2]

In Chap. 1, we mentioned "inventory centralization" uses the law of nature *aggregation reduces variability* to reduce the variability of the business environment, which improves process competencies. In this section, we study inventory centralization from the lens of games in characteristic form, and the discussion adheres to the arguments presented in Kemahlioğlu-Ziya and Bartholdi III (2011).

Consider a supply chain setting with a supplier S and N retailers (indexed as $1, 2, \ldots, n$). The focus is on the single-period problem for a perishable product like apparel. The supplier carries the inventory, acts as a *newsvendor*, and fulfills retailers' orders from the available inventory. Let p, w, h, and p_M denote the wholesale price charged by the supplier, procurement cost per unit for the supplier, holding cost per unit for the supplier, and markup over p at which retailers sell, respectively. Also, let D_i (with d_i being the realization) be the random demand at retailer i ($i \in \{1, 2, \ldots, n\}$) with cdf $F_i(\cdot)$ and pdf $f_i(\cdot)$. We assume that each D_i is independent, and the demand distribution of each retailer is known to the supplier.

Consider the case when the supply chain follows the *decentralized inventory scheme*, as shown in Fig. 8.1, in which the inventory for each retailer is managed

---

[1] It is important to note that the notations used in each section in this chapter are specific to that application.

[2] In this section, the OM actions are on both the *asset side* and the *process side* in the VCAP framework.

© The Author(s), under exclusive license to Springer Nature Singapore Pte Ltd. 2024
R. K. Amit, *Game Theory with Applications in Operations Management*, Springer Texts in Business and Economics, https://doi.org/10.1007/978-981-99-4833-8_8

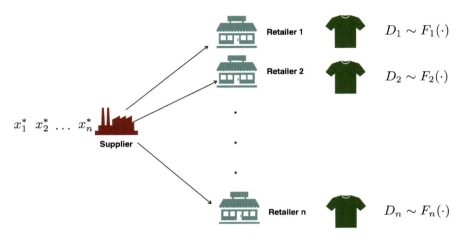

Fig. 8.1 Supply Chain with Decentralized Inventory Scheme

separately, and there is no pooling possible. The supplier's objective function for each retailer in the decentralized inventory scheme is given in Eq. 8.1.

$$\max_{x_i} p\mathbb{E}[\min\{x_i, D_i\}] - h(x_i - \mathbb{E}[\min\{x_i, D_i\}]) - wx_i \qquad (8.1)$$

Using the critical fractile from Eq. 1.3, the optimal inventory carried by the supplier for retailer i is $x_i^* = F_i^{-1}(\frac{p-w}{p+h})$. The total inventory carried by the supplier is $\sum_{i=1}^{N} x_i^*$.

As the supplier bears the inventory risk in the decentralized scheme, she[3] prefers the *pooled inventory scheme* over the decentralized inventory scheme. With pooled independent demands, pooled demand distribution D_c has cdf $F_c(\cdot) = F_1(\cdot) \times \cdots \times F_n(\cdot)$. The pooled inventory scheme is shown in Fig. 8.2.

In the pooled inventory scheme, the supplier sets the inventory level by maximizing the objective function given in Eq. 8.2.

$$\max_{x_S} p\mathbb{E}[\min\{x_S, D_c\}] - h(x_S - \mathbb{E}[\min\{x_S, D_c\}]) - wx_S \qquad (8.2)$$

Using the critical fractile from Eq. 1.3, the optimal inventory carried by the supplier in the pooled scheme is $x_S^* = F_c^{-1}(\frac{p-w}{p+h})$. Due to the pooling effect, $x_S^* \leq \sum_{i=1}^{N} x_i^*$, and the supplier's expected profit in the pooled inventory scheme is higher than the expected profit in the decentralized scheme. That means that while the supplier prefers the pooled scheme, the retailers may be deterred due to the potential decrease in expected sales due to a decrease in inventory allocation.

Using ideas from cooperative games, the supplier proposes the *cooperative pooled inventory scheme*, in which the supplier aims to distribute her excess profits due to pooling among the retailers to adequately compensate for the anticipated decrease in

---

[3] She is aware of the law that aggregation reduces variability.

8.1 Inventory Centralization in Supply Chains

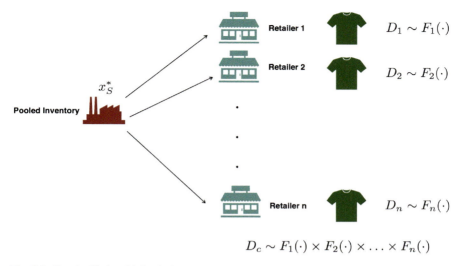

Fig. 8.2 Supply Chain with Pooled Inventory Scheme

sales. Furthermore, it is important to note that the contributions made by each retailer to the supplier's overall gain vary due to each retailer's unique characteristics and circumstances. The allocation method must exhibit fairness and efficiency, hence creating incentives for retailers to join the new scheme. The cooperative pooled inventory scheme leads to a coalition of the supplier and the retailers, as shown in Fig. 8.3.

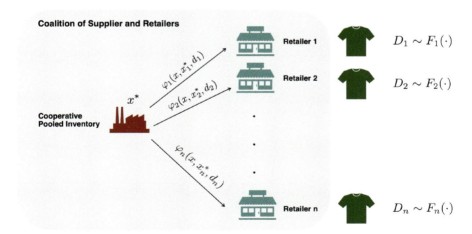

Fig. 8.3 Supply Chain with Cooperative Pooled Inventory Scheme

The cooperative pooled inventory scheme can be modeled as game in characteristic form. Assume for the retailers, the decentralized inventory scheme is the status quo scheme. For realized demand d_i, retailer i's profit is $p_M \min\{x_i^*, d_i\}$ in the decentralized scheme. To incentivize the retailers to participate in the cooperative pooled inventory scheme, the supplier must guarantee that retailer i's profit in the proposed cooperative scheme is at least $p_M \min\{x_i^*, d_i\}$. $\varphi_i(x, x_i^*, d_i) \geq 0$ is the additional profit allocated to retailer i in the event of pooling where x is the supplier's inventory. $\varphi_i(x, x_i^*, d_i)$ is transferable utility (or payoff) by the supplier to retailer i. This is like expectation damages provided by the supplier if she fails to deliver retailer i's preferred order quantity x_i^*. For realized demand d_i, the total profit of retailer i is $p_M \min\{x_i^*, d_i\} + \varphi_i(x, x_i^*, d_i)$. The expected profit of retailer i in the cooperative pooled inventory scheme is given in Eq. 8.3.

$$\Pi_i = p_M \mathbb{E}[\min\{x_i^*, D_i\}] + \mathbb{E}[\varphi_i(x, x_i^*, D_i)] \tag{8.3}$$

Let $\pi_i = p_M \mathbb{E}[\min\{x_i^*, D_i\}]$ and $\phi_i = \mathbb{E}[\varphi_i(x, x_i^*, D_i)]$, then $\Pi_i = \pi_i + \phi_i$.

The cooperative pooled inventory scheme can be represented as game in characteristic form $\langle N \cup S, v \rangle$. The payoff of the grand coalition is given in Eq. 8.4.

$$v(N \cup S) = \mathbb{E}\left[(p + p_M) \min\left\{ x, \sum_{i=1}^{N} D_i \right\} - h \max\left\{ 0, x - \sum_{i=1}^{N} D_i \right\} - wx \right] \tag{8.4}$$

If the grand coalition forms, the expected profit of the supplier is given in Eq. 8.5.

$$\Pi_S = v(N \cup S) - \sum_{i=1}^{N} \Pi_i$$

$$= \mathbb{E}\left[(p + p_M) \min\left\{ x, \sum_{i=1}^{N} D_i \right\} - h \max\left\{ 0, x - \sum_{i=1}^{N} D_i \right\} - wx \right] - \sum_{i=1}^{N} (\pi_i + \phi_i) \tag{8.5}$$

Let π_{S_i} represent the supplier's expected profit because of retailer i in the decentralized inventory scheme, Π_S can be rewritten as

$$\Pi_S = \underbrace{\sum_{i=1}^{N} \pi_{S_i} + \mathbb{E}\left[(p + p_M) \min\left\{ x, \sum_{i=1}^{N} D_i \right\} - h \max\left\{ 0, x - \sum_{i=1}^{N} D_i \right\} - wx \right]}_{a}$$

$$\underbrace{- \sum_{i=1}^{N} (\pi_{S_i} + \pi_i)}_{b} - \underbrace{\sum_{i=1}^{N} \phi_i}_{c} \tag{8.6}$$

8.1 Inventory Centralization in Supply Chains

In Eq. 8.6, $a - b - c$ is the supplier's expected gains when all the retailers switch to the proposed scheme and is denoted as ϕ_S. Π_S can be written as $\Pi_S = \sum_{i=1}^{N} \pi_{S_i} + \phi_S$.

Provided with this updated expected profits formulation, the supplier computes the optimal inventory level x^* such that $\phi_{R_i} \geq 0$ and $\phi_S > 0$, which implies that all agents (supplier S and N retailers) have incentives to join the cooperative pooled inventory scheme and the grand coalition is formed.

Kemahlioğlu-Ziya and Bartholdi III (2011) modify characteristic-form game $\langle N \cup S, v \rangle$ into *inventory pooling game* $\langle N \cup S, w \rangle$ with $w(J) = v(J) - \sum_{i \in J \setminus S} (\pi_{S_i} + \pi_i)$, where $J \subseteq N \cup \{S\}$. $w(J)$ is the expected additional profit for coalition J by pooling the inventory. For any coalition without the supplier ($J \subseteq N$), $w(J) = 0$. Also, the supplier alone cannot generate any additional profit, $w(\{S\}) = 0$. Kemahlioğlu-Ziya and Bartholdi III (2011) show that inventory pooling game $\langle N \cup S, w \rangle$ is a superadditive game (refer Definition 2.5), with the *nonnegative Shapley values* ϕ_i and ϕ_S (refer Sect. 7.3.2) as given in Proposition 8.1. The nonnegative Shapley values ensure the participation of the supplier and the retailers in the cooperative pooled inventory scheme.

Proposition 8.1 (Kemahlioğlu-Ziya & Bartholdi III, 2011) *In inventory pooling game $\langle N \cup S, w \rangle$, the Shapley value allocations to retailer i and the supplier are*

$$\phi_i = \sum_{\substack{J \subseteq \mathcal{N} \cup S \setminus \{i\}: \\ \{S\} \subseteq J, |J| \geq 1}} \frac{(|J|)!(N - |J|)!}{(N+1)!} v(J \cup \{i\}) - \sum_{\substack{J \subseteq \mathcal{N} \cup S \setminus \{i\}: \\ \{S\} \subseteq J, |J| \geq 2}} \frac{(|J|)!(N - |J|)!}{(N+1)!} v(J) - \frac{1}{2}(\pi_{S_i} + \pi_i)$$

$$\phi_S = \sum_{J \subseteq \mathcal{N}: |J| \geq 1} \frac{(|J|)!(N - |J|)!}{(N+1)!} v(J \cup \{S\}) - \frac{1}{2} \sum_{i \in \mathcal{N}} (\pi_{S_i} + \pi_i) \qquad \blacktriangleleft$$

In Sect. 7.3.1, we discussed that the core is one of the fundamental stability-based solution concepts and ensures the stability of the grand coalition. Furthermore, Theorem 7.3 states that the convexity of characteristic-form game is the sufficient condition for the core to be nonempty, and for the Shapley value to be in the core. Inventory pooling games are not necessarily convex, and hence the Shapley value in Proposition 8.1 is not guaranteed to lie in the core, which means the Shapley value allocations in inventory pooling game ensure participation of the supplier and the retailers, but the grand coalition may be unstable. One interesting result is Proposition 8.2, which states that the Shapley value allocation is in the core if all the retailers are identical.

Proposition 8.2 (Kemahlioğlu-Ziya & Bartholdi III, 2011) *The Shapley value allocation is in the core with identical retailers.* $\qquad \blacktriangleleft$

8.2 Service Systems[4]

We discussed the law of nature "aggregation reduces variability" in Chap. 1. We saw the application of the law in Sect. 8.1, where the efficacy of using the Shapley values for consolidating inventory of a homogeneous good within supply chains was discussed. The aggregation of inventory of the homogeneous good leads to *economies of scale*. In this section, we represent and analyze *economies of scope*[5] when nonidentical products are aggregated as characteristic-form game.

Economies of scope is illustrated in Fig. 8.4. An information technology-based manufacturing firm produces products such as computers, laptops, monitors, mobile, and landline phones. The firm also provides a comprehensive after-sales service, ensuring customer satisfaction through the deployment of highly skilled staff. Due to inherent variability in demand and service times of after-sales services, the firm can gain by aggregating (or pooling) the service types. This can be achieved by training service personnel in multitasking, as shown in Fig. 8.4. This aggregation reduces variability in the system that, in accordance with Little's law and Kingman's formula (refer Chap. 1), reduces cycle time and queue length. If each service is provided by an independent service provider within the firm, it is imperative that the gain is allocated in an equitable manner. The situation is modeled as game in characteristic form by Anily and Haviv (2010), which is the basis of our further discussion.

Consider N service providers, indexed by $i = \{1, 2, .., n\}$, where each service provider independently serves its customers. Each server is characterized by Poisson arrival rate λ_i and exponential service rate $\xi_i > \lambda_i$ when working individually. However, the actual service rate μ_i depends on the arrival rate to the system and assumes $\lambda_i < \mu_i$ for stability. Also, suppose that μ_i is the weighted geometric mean of λ_i and ξ_i such that $\mu_i = \lambda_i^{1-\alpha}\xi_i^{\alpha}$, for a given $\alpha \in (0, 1]$. Regard server i's utilization level as $\rho_i = \lambda_i/\mu_i = (\lambda_i/\xi_i)^{\alpha}$.

As discussed earlier, the service providers can pool the services to reduce variability. $S \subseteq N$ is a coalition of service providers who pool their services. Coalition S has arrival rate $\lambda(S) = \sum_{i \in S} \lambda_i$, service rate $\xi(S) = \sum_{i \in S} \xi_i$, and actual service time is $\mu(S) = \lambda(S)^{1-\alpha}\xi(S)^{\alpha}$. Given these parameters, the expected number of customers in steady state $v(S)$, is given in Eq. 8.7.

$$v(S) = \frac{\rho(S)}{1 - \rho(S)} = \frac{\lambda(S)}{\mu(S) - \lambda(S)} = \frac{\lambda(S)^{\alpha}}{\xi(S)^{\alpha} - \lambda(S)^{\alpha}}, \tag{8.7}$$
$$v(\varnothing) = 0, v(\{0\}) = \infty \text{ and } v(\{n + 1\}) = 0$$

$\langle N, v \rangle$ is characteristic-form game that models cooperation in the service system. It can be verified that $\langle N, v \rangle$ is subadditive, which means that the grand coalition is efficient. However, it is not a concave game (refer to convex games in Definition 7.3);

---

[4] In this section, the OM actions are on the *process side* in the VCAP framework.

[5] We recommend (Panzar & Willig, 1981) for economies of scope.

8.2 Service Systems

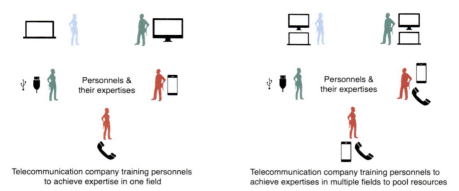

Fig. 8.4 Cooperation in Service Systems

Table 8.1 Characteristic functions

S	{1}	{2}	{3}	{1, 2}	{1, 3}	{2, 3}	{1, 2, 3}
$v(S)$	9	1	1/9	7/3	1	3/7	1
$w(S)$	1	3/7	1/9	1	1	3/7	1

hence, the nonemptiness of the core for a stable grand coalition is not guaranteed. Also, the Shapley value allocations may not lie in the core.

To achieve concavity, a new characteristic function $w(S) = \min_{T|S \subseteq T \subseteq N} v(T)$ can be defined, where $w(S)$ is the cost of the best coalition that also contains service providers in S. Anily and Haviv (2010) show that characteristic-form game $\langle N, w \rangle$ is concave and *balanced*. Balanced games have nonempty core (Theorem 7.1). Also convex games (concave games in this section) have nonempty core and the Shapley value is in the core (Theorem 7.3). This means $\langle N, w \rangle$ has nonempty core, with the Shapley value is in the core. This is illustrated using the following numerical example. Let $\alpha = 1$ and $N = \{1, 2, 3\}$. Suppose $\{\lambda_1, \lambda_2, \lambda_3\} = \{9, 5, 1\}$ and $\{\xi_1, \xi_2, \xi_3\} = \{10, 10, 10\}$, which yields $\{\rho_1, \rho_2, \rho_3\} = \{0.9, 0.5, 0.1\}$. The characteristic functions are in Table 8.1. The Shapley value vector for $\langle N, v \rangle$ is $(\frac{673}{189}, \frac{-137}{189}, \frac{-347}{189})$, which is not in the core, whereas the Shapley value vector for $\langle N, w \rangle$ is $(\frac{145}{189}, \frac{37}{189}, \frac{1}{27})$, which is in the core.

The negative Shapley values for $\langle N, v \rangle$ in the previous example can be explained as payments by service providers to persuade faster service provider, whose processing time is low and the server is mostly idle, to join their coalition, which brings tremendous improvement in service times, characterize the payments with negative entries. Hence, it is important to characterize the core allocations with negative payments. Let the collection of best service providers be $BN = \{i : v(N) < v(N_{-i})\}$. Theorem 8.1 certifies that a negative payment is possible only when there are best service providers in the system, whereas Theorem 8.2 establishes the bounds on the payment to the best service provider.

150   8 Games in Characteristic Form: Applications in OM

Theorem 8.1 (Anily & Haviv, 2010) *If $i \notin BN$, then the core allocation for $\langle N, v \rangle$ with negative payment to i does not exist.*   ◄

Theorem 8.2 (Anily & Haviv, 2010) *For any $i \in BN$, the payment to i in any core allocation exceeds $v(N) - v(N_{-i})$, and there exist infinitely many such core allocations.*   ◄

Guo et al. (2013), Bendel and Haviv (2018), and Karsten et al. (2015) extend the research of Anily and Haviv (2010).

8.3 Towards Sustainability

In Example 7.2, we discussed the *recycling game* to model cooperation among automobile manufacturers to achieve *extended producer responsibility* (EPR) goals for sustainability. This section discusses further applications of cooperative game theory in attaining environmentally sustainable goals.

8.3.1 Recycling Coalitions[6]

Extended Producer Responsibility (EPR) is a setting that has gained significant attention in the field of cooperative game theory. This concept has been widely embraced by numerous governments in recent times. In a nutshell, EPR mandates that manufacturers assume responsibility for the appropriate disposal of products following their use by consumers. Before the implementation of EPR, the responsibility for managing end-of-life items rested mostly with governments, while producers showed little concern for retrieving and recycling discarded products.

The implementation of EPR entails a transfer of the recycling responsibility from governmental entities to producers. This phenomenon leads to manufacturers exhibiting awareness during production and exerting efforts post-consumption, as failure to comply with EPR might incur significant expenses. The ability to monitor end-of-life items is facilitated by contemporary technology. However, the challenge in recycling primarily stems from the variability of materials and the substantial fixed costs involved.

The establishment of recycling coalitions, when producers get together to pool their resources, may first appear to be a promising approach. However, empirical data indicates that such coalitions tend to exhibit high instability. In the year 2020, Samsung, Toshiba, and Panasonic were affiliated with MRM, an organization located in the United States. In recent times, Samsung has terminated its affiliation with MRM, but Toshiba and Panasonic continue to maintain their status as MRM clients. The

---

[6] In this section, the OM actions are on both the *asset side* and the *process side* in the VCAP framework.

8.3 Towards Sustainability

inquiry at hand pertains to the factors influencing these alliances and the feasibility of establishing a sustainable recycling network.

The primary obstacle impeding the recycling of used items is the substantial financial burden associated with material separation, mostly attributable to the diverse composition of materials and the costly infrastructure necessary for this process. The key factor influencing the cost of recycling units is the presence of material heterogeneity. There is a positive correlation between the degree of material heterogeneity and the associated cost of separation. When the cost of raw materials is not high, corporations tend to choose the option of "grinding the product" for the purpose of recycling. However, the presence of important reusable components typically poses challenges for firms when it comes to recycling, as the process of separating these components can be labor-intensive.

As a result of intense competition, manufacturers actively seek differentiation techniques in their production processes in order to gain a competitive advantage. This, in turn, leads to increased material heterogeneity at the macro-level. One potential strategy is implementing recycling practices at the level of individual producers, which is commonly referred to as *individual producer responsibility* (IPR) (Atasu et al., 2010). In order to achieve coherence in the recycling process of automobiles, it is important to establish distinct recycling facilities for various components such as tires, seats, batteries, and other relevant materials. This approach aims to mitigate the discrepancies in recycling these diverse components. Nevertheless, the task of establishing and sustaining a specialized recycling system inside the realm of producers poses significant challenges and financial burdens.

The potential advantages derived from recycling may not be sufficient to achieve a point of equilibrium due to the scarcity of available used items. In such circumstances, it is imperative for individual producers to engage in collective recycling efforts in order to reach a point of equilibrium and reap the benefits associated with economies of scale. The approach is referred to is commonly known as *collective producer responsibility* (CPR) (Gui et al., 2015). Both methodologies possess inherent drawbacks, and governmental entities are actively engaged in the administration of recycling endeavors.

The implementation of EPR by the European Union (EU) in 2002 resulted in the transfer of the government's responsibility for collection, recycling, and recovery to the producers. The legislation was enacted and then adopted on a global scale. Local governments within the European Union (EU) enforce the mandatory participation of producers in various recycling initiatives, such as **European Recycling Platform**. Despite the absence of federal approval for extended producer responsibility legislation in the United States, Texas implemented take-back regulations in 2007 and 2011 for computer trash and television waste, respectively, with the aim of promoting recycling practices. Washington State and New York State have comparable legal frameworks.

The prevailing political circumstances provide producers the liberty to participate in, withdraw from, or initiate a recycling initiative. In 2007, Panasonic was affiliated with **EcologyNet Europe** as a member. In a subsequent development in the year 2020, it became a member of **REPIC**, an association that includes LG, Sharp, and

Toshiba. Furthermore, there is a growing imperative to prioritize producing and consuming environmentally friendly products, as consumers are becoming cognizant of the importance of utilizing eco-friendly alternatives.

The term *recycling coalition* pertains to the participants involved in a recycling initiative, which can vary from a single entity to several entities. On the other hand, a *recycling network* encompasses all the producers and their respective coalitions inside the recycling program. The establishment of extensive recycling coalitions facilitates the exchange of technology and resources among participating members, reducing fixed costs. However, the presence of material heterogeneity is substantial, resulting in a labor-intensive disassembly process, which subsequently increases the unit cost of recycling. It can be inferred that coalitions that are either too large or too small in size are inefficient in recycling. Moreover, available information indicates that the majority of coalitions focused on recycling exhibit a lack of stability. Tian et al. (2020) study the impact of recycling costs on the establishment and sustainability of a recycling network and use *farsighted stability* to ensure the stability of the recycling network. The following discussion is based on Tian et al. (2020).

Given EPR regulations, producers have to bear the cost of returning and recycling the used goods once they become obsolete. The recycling capabilities are considered while determining the initial manufacturing levels. The mode or way of recycling is flexible as producers have the liberty to choose, and coalition formation is assumed to be *endogenous*.

Let $\mathcal{N} = \{1, ..., N\}$ be the set of identical producers. $\mathcal{A} = \{A_1, \ldots, A_j \ldots, A_J\}$, where $A_j \subseteq \mathcal{N}$ is a coalition structure if the coalitions are non-intersecting and their union is $\mathcal{N}$. Also, $\mathbf{n} = (n_1, ..., n_J)$ is a recycling network, with $n_1 \leq n_2 \leq ... \leq n_J$ where $n_j = |A_j|$. To understand the terminology, consider the following illustration. Let there be 10 producers, and they form two coalitions $A_1 = \{1, 2, 5, 8\}$ and $A_2 = \{3, 4, 6, 7, 9, 10\}$. The coalition structure is $\mathcal{A} = \{\{1, 2, 5, 8\}, \{1, 3, 4, 6, 7, 9, 10\}\}$ and the recycling network is $\mathbf{n} = (4, 6)$ indicating that there are two coalitions, one with four producers and the other with six.

Tian et al. (2020) consider fixed cost and variable cost to characterize the recycling costs. The cost structure implicitly includes activities like tracking, collection, disassembling, and disposal. Assume that the third-party recyclers are sought for recycling, wherein they charge a fixed cost (K) independent of coalition size and material heterogeneity, whereas variable cost depends on both. If $\bar{c}(n_j)$ is the unit recycling cost for coalition A_j, then the variable cost for coalition A_j is given in Eq. 8.8.

$$c(A_j) = \bar{c}(n_j)x \sum_{i \in A_j} q_i \qquad (8.8)$$

where q_i is the quantity to be recycled for producer i and $x \in [0, 1]$ denotes the level of material heterogeneity. $x = 0$ indicates completely homogeneous products and $x = 1$ denotes completely heterogeneous products. $x = 0 \implies c(A_j) = 0$, which

8.3 Towards Sustainability

means that only the fixed cost is incurred with completely homogeneous products. The total recycling cost $C(\mathbf{q}, \mathbf{n})$ for the coalition network $\mathbf{n}$ is given in Eq. 8.9.

$$C(\mathbf{q}, \mathbf{n}) = \sum_{j=1}^{J} (c(A_j) + K) \tag{8.9}$$

where $\mathbf{q} = \{q_1, ..., q_N\}$ is the vector indicating quantity for each producer. For producer $i \in A_j$, recycling cost depends on the quantity q_i and its share of the fixed cost, and is written as $\bar{c}(n_j) x q_i + \frac{K}{n_j}$. Under EPR regime, producers decide their quantities based on the recycling capabilities to maximize respective payoff $\pi_i(\mathbf{q}, n_j)$. Assuming Cournot competition among the producers, equilibrium outcome $q_i^*(\mathbf{n})$ is identical for all producers, as they are assumed to be identical.

For notational convenience, let $\pi_i(\mathbf{q}^*, n_j) = \pi(\mathbf{n}, n_j)$. Such a representation is compact and allows one to denote the payoff of identical producers in the recycling network. For instance, with five producers forming *two coalitions* with two and three producers ($\mathbf{n} = (2, 3)$), $\pi((2, 3), 2)$ and $\pi((2, 3), 3)$ are the payoffs of any producer in two-producer and three-producer coalitions, respectively. Now, the producers' decisions are dependent on the endogenous coalition formation, and the optimization problem is $\max_{\mathbf{n}} \pi(\mathbf{n}, n_j)$. Since this optimization involves the interaction of producers, stability-based solution concepts are relevant in computing the solution of the optimization problem. *Farsighted stability* is considered here as the stability-based solution concept.

A recycling coalition is *unstable* if players are better off deviating from the coalition. Consider a setting with four producers. The producers' preferences are $\pi((1, 3), 1) \succ \pi((4), 4) \succ \pi((1, 1, 1, 1), 1) \succ \pi((1, 3), 3) \succ \pi((A), n_j), \forall A \notin \{(4), (1, 3), (1, 1, 1, 1)\}$. From the preferences, we can infer that the grand coalition is not stable as a player has an incentive to leave the grand coalition and act alone ($\pi((1, 3), 1) \succ \pi((4), 4)$). If the player leaves the grand coalition, then the remaining players are also better off at dismantling the coalition and acting alone as $\pi((1, 1, 1, 1), 1) \succ \pi((1, 3), 3)$. Again, $\pi((4), 4) \succ \pi((1, 1, 1, 1), 1)$, the players prefer to form the grand coalition, which is unstable. Hence, a stable coalitional network does not exist if we follow myopic reasoning.

A careful reader would have noticed a pattern in the example presented above. For instance, after a sequence of deviations from the grand coalition, the grand coalition is again attained, that is, $(4) \rightarrow (1, 3) \rightarrow (1, 1, 2) \rightarrow (1, 1, 1, 1) \rightarrow (4)$. Producers know that the deviation from the grand coalition leads to a sequence of further deviations that end up giving a payoff less than the grand coalition. In some sense, (4) can be regarded as a stable network. This idea of attaining stability in sequential reactive steps is called *farsighted stability*, where the players consider the possible reactions of other players before deviating and are less concerned with the immediate consequences. We recommend (Chwe, 1994) for further reading on farsighted coalition stability.

Tian et al. (2020) mention that the stability of a coalition network is determined by its level of heterogeneity and fixed cost, specifically the unit recycling cost. This

8 Games in Characteristic Form: Applications in OM

stability is observed across a wide range of conditions, indicating the farsighted approach. Tian et al. (2020) also examine operationalizing the coalitional network by means of taxes and subsidies in order to maintain stability.

8.3.2 GREEN Game[7]

This section provides an overview of the application of cooperative game theory in controlling greenhouse gas (GHG) emissions. Gopalakrishnan et al. (2021) consider a supply chain wherein several suppliers, manufacturers, and assemblers cooperate to produce the finished product. Let the firms be denoted as $N = \{1, 2, ..., n\}$ and processes as $M = \{1, 2, ..., m\}$ with corresponding carbon footprints as $\mathbf{f} = \{f_1, f_2, ..., f_m\}$. f_j for process j can be influenced by a subset of firms $N^j \subseteq N$, directly or indirectly. For instance, transportation decisions depend on the manufacturer's operational decisions. Regardless of direct or indirect influence, an entry in the responsibility matrix $\mathbf{B} = (b_{ij})$ is 1 if $i \in N^j$, and 0 otherwise.

Assume that the supply chain leader incurs a penalty $p^S > 0$ per ton of carbon emissions in the supply chain. The supply chain leader strives to identify a pollution-allocation rule to mitigate the total pollution $\sum_{j \in M} f_j$ by leveraging $\mathbf{B}$ and incentivizing firms to take appropriate actions. Formally, the leader devises *pollution-allocation rule* $\mathbf{Œ}(\mathbf{B}, \mathbf{f})$ that specifies the total pollution responsibility for firm i, $\phi_i = \phi_i(\mathbf{B}, \mathbf{f})$ such that $\sum_{i=1}^n \phi_i = \sum_{j=1}^m f_j$. The model formulation involves two stages:

- **Stage 1**: Each firm $i \in N^j$ exerts efforts to reduce f_j.
- **Stage 2**: The total pollution $\sum_{j \in M} f_j$ is allocated in the supply chain based on the pollution-allocation rule $\mathbf{Œ}(\mathbf{B}, \mathbf{f})$, and the corresponding carbon penalty $p^S \phi_i$ is collected.

Note that the decisions in Stage 1 and 2 are intertwined as the carbon penalties in Stage 2 motivate the firms to exert more effort. The characteristic function for Stage 2 can be constructed to capture the pollution responsible for the firms that form a coalition. For $S \subseteq N$, characteristic function $c_G(S)$ is given in Eq. 8.10.

$$c_G(S) = \sum_{j \in M} f_j b_{S,j} \tag{8.10}$$

where $b_{S,j} = 1$ if $b_{i,j} = 1$ for any $i \in S$, and 0 otherwise. Characteristic-form game $\langle N, c_G \rangle$ is called *GHG Responsibility-Emissions and Environment game* (**GREEN game**). The representation of GREEN game is compact to make the solution tractable and efficiently computable. c_G satisfies $c(S \cup \{i\}) - c(S) \le c(Q \cup \{i\}) - c(Q), \forall i \notin S$ and $Q \subseteq S \subseteq N$, hence $\langle N, c_G \rangle$ is concave (refer convex games in Definition 7.3) . This is the result in Proposition 8.3.

---

[7] In this section, the OM actions are on the *process side* in the VCAP framework.

8.4 Logistics Networks

Proposition 8.3 (Gopalakrishnan et al., 2021) *GREEN game is concave.* ◀

Proposition 8.3 ensures that the core is nonempty and Shapley value lies in the core (Theorem 7.3). The Shapley value of $\langle N, c_G \rangle$ is computed using the result in Proposition 8.4.

Proposition 8.4 (Gopalakrishnan et al., 2021) *Emission f_j allocated equally among all firms in N^j is the Shapley value of $\langle N, c_G \rangle$.* ◀

Here, we do not discuss the intrinsic mechanisms underlying abutment strategies in Stage 1. Gopalakrishnan et al. (2021) compare the abatement strategies of two apparel suppliers of Walmart, namely, Levi Strauss and Nautica.

8.4 Logistics Networks[8]

In Sect. 4.4, we mentioned that numerous technologies have been instrumental in reducing transaction costs (refer Fig. 4.9), which led to the emergence of supply chains. With increasing globalization, supply chains are becoming complex and long, with longer distances, higher lead times, and more players. Furthermore, products are also becoming more complex and have a wide variety. These challenges in supply chains are illustrated in Fig. 8.5. However, in a global supply chain, the source

Fig. 8.5 Supply Chain Complexity

location may not be directly connected with the destination location. Cost advantages can be achieved by shipping the goods to an intermediate location before reaching the final destination through *logistic networks*. Logistic networks require *transshipment*—movement of goods from one location in the supply chain to another, where the goods are unloaded to undergo processing or temporarily stored before moving on to the next location. A simple transshipment setting is shown in Fig. 8.6. Chiou

---

[8] In this section, the OM actions are on the *process side* in the VCAP framework.

(2008) provide a review of transshipment problems in supply chains. As evident from Fig. 8.6, each route/mode in logistic networks may be controlled or owned by different agents. Cooperative game theory is used to model cooperation among the network agents to achieve the network goals. We follow Reyes (2005) to discuss the usage of games in characteristic form in a transshipment setting.

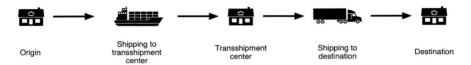

Fig. 8.6 Transshipment Process

Transshipment problem lies in a broader class of games called *flow games*. A flow game aims to allocate the number of units of a good that must travel through a given arc in the network. A flow game is represented using a *directed graph* with a set of nodes represented by V and a set of connected links called A. The directed graph is denoted as $G = (V, A)$. The flow begins through an entry node called the *source* and an exit node called the *sink*. The sink node necessitates a specific demand that must be satisfied, while the source node provides the supply. The players of the flow game have control over the links in the game. The aim of the flow game is to allocate the goods that flow through each link by establishing cooperation among the players. The system of vertices V is connected by links a_{ij}, where $i \neq j$ denotes the movement direction from node i to node j. Each link has a capacity of c_{ij} units. x_{ij} is the decision variable that allocates quantity on link a_{ij}. An individual player can control a link or a group of links. The players can also form coalitions, allowing the coalition to control a set of links.

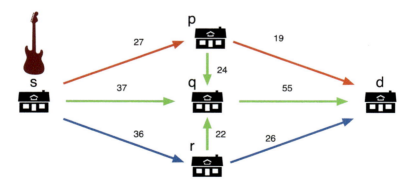

Fig. 8.7 Logistics network

Let us consider a logistic network for transporting as shown in Fig. 8.7. The network has one supplier (s), three distributors (p, q, and r), and one destination

8.4 Logistics Networks

Table 8.2 Characteristic function values of possible coalitions

Coalition	Links available	Value of the sub-network
1	$(s, p), (p, d)$	$v(1) = 19$
2	$(s, q), (p, q), (r, q), (q, d)$	$v(2) = 37$
3	$(s, r), (r, d)$	$v(3) = 26$
1, 2	$(s, p), (p, d), (s, q), (p, q), (r, q), (q, d)$	$v(12) = 64$
1, 3	$(s, p), (p, d), (s, r), (r, d)$	$v(13) = 45$
2, 3	$(s, q), (p, q), (r, q), (q, d), (s, r), (r, d)$	$v(23) = 73$
1, 2, 3	$(s, p), (p, d), (s, q), (p, q), (r, q), (q, d), (s, r), (r, d)$	$v(123) = 100$

(d), which is the retailer. The goods[9] begin their journey from the supplier and transshipped to either of the three distributors and finally end up with the retailer. The total number of units to be transported is 100. The links are owned by three players: P1, P2, and P3. P1 owns the upper links (shown in red color) (s, p) and (p, d). P2 owns the middle links (shown in green color) (s, q), (p, q), (r, q), and (q, d). P3 owns the remaining lower links (shown in blue color) (s, r) and (r, d). The capacity of each link is also given, as shown in Fig. 8.7.

The logistic network can be modeled as flow game in characteristic form $\langle N, v \rangle$, where N is the set of players, $v(S)$ is the characteristic function that gives the maximum flow for coalition S where $S \subseteq N$. In other words, $v(S)$ is the maximum number of units that can be shipped through the available sub-network involving players in coalition S. With $N = \{P1, P2, P3\}$, Table 8.2 shows the characteristic function value of each possible nonempty coalition. It can be verified that the game is superadditive (refer Eq. 2.5), which means cooperation among the players increases the maximum volume that can be shipped through the sub-network.

As mentioned earlier, 100 units of the good need to be transported, and let us assume that the price per unit transported is \$1. The total revenue generated by transporting 100 units is \$100. Assume P1, P2, and P3 are aware of the Shapley value as a fairness-based solution concept (refer Sect. 7.3.2), and use it to share the revenue generated by cooperatively transporting the good. Using the Shapley value given in Theorem 7.2, the Shapley values for P1, P2, and P3 are computed in Eq. 8.11. The Shapley value measures a player's importance and recommends sharing the revenue generated based on each player's importance in the cooperative game. For example, the links owned by P1 can transport a maximum of 19 units; however, the Shapley value ϕ_1 is 23, which captures the contribution of P1 to the network.

---

[9] In this case, the goods are electric guitars that need to be safely transported. There was an incident when musician Dave Carroll claimed that his guitar was broken while in United Airlines' custody. Dave Carroll and his band, "Sons of Maxwell", composed a protest song "United Breaks Guitars". The song became an instant hit on YouTube, and even impacted United Airlines' stock prices.

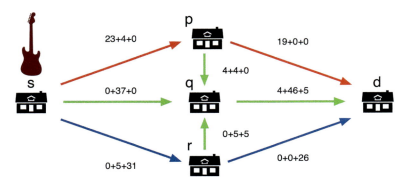

Fig. 8.8 Logistics network with Shapley Value

$$\phi_1 = \frac{1}{3}[v(1) - v(0)] + \frac{1}{6}[v(12) - v(2)] + \frac{1}{6}[v(13) - v(3)] + \frac{1}{3}[v(123) - v(23)]$$
$$= 23$$
$$\phi_2 = \frac{1}{3}[v(2) - v(0)] + \frac{1}{6}[v(12) - v(1)] + \frac{1}{6}[v(23) - v(3)] + \frac{1}{3}[v(123) - v(13)]$$
$$= 46$$
$$\phi_3 = \frac{1}{3}[v(3) - v(0)] + \frac{1}{6}[v(13) - v(3)] + \frac{1}{6}[v(23) - v(2)] + \frac{1}{3}[v(123) - v(12)]$$
$$= 31$$

(8.11)

Using the Shapley value, the allocation of quantity to each link is shown in Fig. 8.8. In the cooperative network, to transport 100 units, P1 and P3 transport some of the units for P2 when link (s, q) has limited capacity, and P2 accepts four units from P1 and five units from P3 to meet the capacity constraints in links (p, d) and (r, d). Furthermore, the Shapley value payoff vector (ϕ_1, ϕ_2, ϕ_3) also lies in the core of the game, as shown in Eq. 8.12, which means that revenue sharing based on the Shapley values ensures the stability of the logistics network.

$$\begin{aligned}
\phi_1 &= 23 > 19 = v(1), \\
\phi_2 &= 46 > 37 = v(2), \\
\phi_3 &= 31 > 26 = v(3), \\
\phi_1 + \phi_2 &= 69 > 64 = v(11), \\
\phi_2 + \phi_3 &= 77 > 73 = v(23), \\
\phi_1 + \phi_3 &= 54 > 45 = v(13), \\
\phi_1 + \phi_2 + \phi_3 &= 100 = 100 = v(123)
\end{aligned}$$

(8.12)

8.5 SHAP Algorithm

We recommend some papers that use cooperative games in logistic networks. Hafezalkotob and Makui (2015) apply the concepts from cooperative game theory to solve the maximal-flow problems, which aim to transport the maximum amount of flow from source to sink, given the capacities of the links. Ergün et al. (2021) demonstrate another application in emergency logistics planning to coordinate logistic support for relief operations. Rudramoorthi and Amit (2023) propose a new class of games called the *repositioning games* that can be used for repositioning ambulances to improve their coverage.

8.5 SHAP Algorithm[10]

We are in the midst of *artificial intelligence* (AI) spring. *Machine learning* (ML) is driving this revolution, and to capture intricate patterns and complexity in data, ML models are becoming increasingly complex. The quest for higher prediction accuracy drives the adoption of complex models (Lundberg & Lee, 2017). As AI and ML find applications across diverse fields, the focus must be on enhancing the *interpretability* and, ultimately, *explainability* of these black box models to build trust and facilitate meaningful decision-making.

Interpretability refers to the extent to which the models capture the cause and effect. As shown in Fig. 8.9, simple linear models, like linear regression, are inherently interpretable, making understanding how individual predictors influence the outcome easy, even if their predictive accuracy might not be exceptional. Explainability refers to the justifications or reasons for the model's decisions. This concept is the predictability of the output of a complex model by a simpler model. Explainability has multifaceted applications for the generated AI models with wide applicability in various fields such as economics, social science, and philosophy. Interpretability and explainability are two important criteria for understanding the model and the predictions. Both are important when it comes to accurately modeling the relationship between the dependent and independent variables. Explainable models are part of the burgeoning field of *explainable* AI (XAI). We recommend (Russell & Norvig, 2020) for further discussion on explainable AI. In this section, we focus on SHAP (SHapley Additive exPlanations), a framework for explaining the output of machine learning models by assigning a value to each feature (input) that indicates its contribution to the model's prediction. It is a powerful concept based on the Shapley value (refer Sect.7.3.2).

---

[10] In this section, the OM actions are on the *process side* in the VCAP framework.

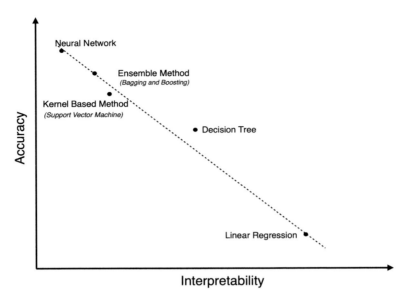

Fig. 8.9 Interpretability and Accuracy

8.5.1 Explanation Models

An original complex model f can be explained easily using a simple explainable model g. The explanation model uses simplified binary input $\mathbf{x}' \in \{0, 1\}^M$ that maps to the original inputs $\mathbf{x} \in \mathbb{R}^M$ through a mapping function $\mathbf{x} = h_{\mathbf{x}}(\mathbf{x}')$. M is the number of feature inputs. The explanation models try to ensure $g(\mathbf{x}') = f(h_{\mathbf{x}}(\mathbf{x}'))$. *Additive feature attribution models* are explanation models with linear function of simplified binary input $\mathbf{x}'$, as shown in Eq. 8.13, with x'_i is the ith component of $\mathbf{x}'$ and $\phi_i \in \mathbb{R}$.

$$g(\mathbf{x}') = \phi_0 + \sum_{i=1}^{M} \phi_i x'_i \qquad (8.13)$$

Two concepts are important to discuss the explainability of a model: *feature importance* and *feature attribution*. Feature importance quantifies the extent to which a feature is responsible for constructing the model, whereas feature attribution refers to the marginal effect of a feature on the model prediction. SHAP algorithm is in the class of additive feature attribution models, and considers both feature importance and feature attribution. It explains the model by allocating the prediction value to each feature according to the marginal contribution. The method decomposes the prediction value $f(\mathbf{x})$ into the contribution of each feature j. This is similar to the Shapley value, where the payoff is divided fairly among the players according to

8.5 SHAP Algorithm 161

their contribution. We can represent the SHAP value $\phi_j(f, \mathbf{x})$ of feature j, as given in Eq. 8.14.

$$\phi_j(f, \mathbf{x}) = \sum_{S \subseteq F \setminus \{j\}} \frac{|S|!(N - |S| - 1)!}{N!} \left[f_x(S \cup \{j\}) - f_x(S) \right] \quad (8.14)$$

where F is the set of all features with cardinality N. S is the subset of features $F \setminus \{j\}$. $f_x(S)$ is the model's predicted value based on the subset S features. $[f_x(S \cup \{j\}) - f_x(S)]$ is the marginal contribution of the feature j when added to feature set S. Similar to the axiomatization of the Shapley value discussed in Sect. 7.3.2, Lundberg and Lee (2017) provide three axioms for the SHAP values,[11] which are

- **Local accuracy**: The sum of all feature attributions is equal to the value of the prediction; $f(\mathbf{x}) = \phi_0(f) + \sum_{j=1}^{N} \phi_j(f, \mathbf{x})$. This is similar to the *efficiency* axiom of Shapley value.
- **Missingness**: The absence of a feature results in no attribution; if $f_x(S \cup \{j\}) = f_x(S)$, then $\phi_j(f, \mathbf{x}) = 0$. This is similar to the *dummy player* axiom of the Shapley value.
- **Consistency**: An increase in the impact of a feature always results in increasing the attribution of that feature; for two models (functional relationship) f and f', if $f'_x(S) - f'_x(S \setminus j) \geq f_x(S) - f_x(S \setminus j) \ \forall S \in F$, then $\phi_j(f', \mathbf{x}) \geq \phi_j(f, \mathbf{x})$. This is the similar to the *monotonicity* axiom of the Shapley value.

Theorem 8.3 (Lundberg & Lee, 2017) *The SHAP value is the only additive feature attribution model that satisfies the above three axioms.* ◄

8.5.2 Explanation Models in Quality Modeling

For semiconductor manufacturing at Hitachi ABB, Senoner et al. (2022) present a study that uses explanation models to improve the quality parameters. Quality improvement has a strong connection with statistical techniques where the objective is to find and eliminate the source for the quality variation (Schmenner & Swink, 1998; Taguchi, 1986). This can be achieved by capturing the relationship between the attributes of a production process and process quality. The SHAP algorithm can be useful for finding the variation source using feature attribution and feature importance. To simplify this, let us consider a manufacturing process that generates data on production parameters x_j, with $j \in \{1, \ldots, N\}$, and the process quality y. An observation i, with $i \in \{1, \ldots, M\}$, gives $(x^{(i)}, y^{(i)})$ as data point. $x_j^{(i)}$ is the value of parameter x_j at i^{th} observation. As evident, the total number of parameters is N, and the total number of observations is M. For each process $k \in \{1, \ldots, K\}$,

---

[11] These axioms are based on another axiomatization of the Shapley value by Young (1985b).

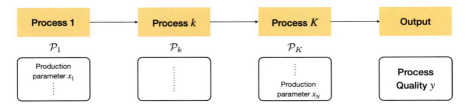

Fig. 8.10 Overview of the Production System. Adapted from Senoner et al. (2022)

process specification $\mathcal{P}_k \subseteq \bigcup_j^N x_j$ specifies the production parameters for process k. It should be noted that a production parameter is unique to a process. Figure 8.10 depicts the entire manufacturing process system.

Manufacturers typically adopt a two-step approach to inform decisions regarding quality improvement. In the initial step, their focus is on prioritizing processes that have the most substantial impact on the overall quality variation of the process. Subsequently, during the second step, they identify potential enhancement measures for the prioritized processes and carefully choose actions with the greatest potential for enhancing quality. The nonlinear modeling approach is suitable for this as there is a possibility that a parameter $x_{j'}$ of a process k' can interact with a production parameter $x_{j''}$ from another process k''. The two-step approach discussed above takes initial input as the historical manufacturing data $\{(x^{(i)}, y^{(i)})\}_{i=1}^M$ and produces the nonlinear model f for the production parameter inputs $x^{(i)}$ and process quality $y^{(i)}$. This is called the phase of *learning of metamodel* where nonlinear relationships are identified using models like tree ensembles or deep neural networks. This model explains the underlying relationship between various production parameters and the process quality. The SHAP value captures this effect by estimating the quality change when a particular production parameter is omitted. SHAP value explains the model f locally at each observation level i. This is formally written as

$$f(x^{(i)}) = \phi_0 + \sum_{j=1}^N \phi_j^{(i)}$$

Here ϕ_0 corresponds to the expected value $\mathbb{E}[f(x)]$ and $\phi_j^{(i)} \in \mathbb{R}$ is the SHAP value of the production parameter $x_j^{(i)}$. The SHAP value for an observation of a production parameter calculates the deviation of the process quality from the expected value. A positive value increases the quality, and a negative value does the opposite. The absolute value signifies the quantum of the change in the expected process quality. This computation is repeated for all observations, leading to the calculation of feature attribution $\phi_j^{(i)}$ for each parameter $j \in \{1, \ldots, N\}$ and each observation $i \in \{1, \ldots, M\}$.

Process importance starts with the above calculation by aggregating the feature attribution across all observations to determine the feature attribution at the production parameter level. Aggregation is done for the absolute values, and the mean

8.6 Aumann–Shapley Pricing

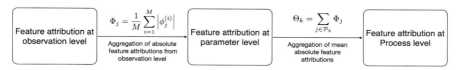

Fig. 8.11 Overview of the Process Importance Calculation Method

absolute feature attribution is obtained. Absolute values are selected to deal with the cancellation effect of negative feature attribution. After obtaining the feature attribution at the parameter level, further aggregation of the mean absolute feature is done to translate the importance to the process level. Figure 8.11 illustrates the three-step approach to obtain process importance from the observation level SHAP values.

This measure of process importance provides a meaningful estimate because of the local accuracy property of the SHAP value. Furthermore, it provides the interaction effect as feature attributions are calculated over all possible subsets of production parameters. Using the process importance calculation depicted above, the most significant process is obtained by the following method:

$$p^* \in \operatorname*{argmax}_{p \in \{0,1\}^K} \sum_{k=1}^{K} p_k \Theta_k \quad \text{s.t.} \quad \sum_{k=1}^{K} p_k \leq p_{\max}$$

$p^* \in \{0, 1\}^K$ indicates which processes should be prioritized for improvement actions. The constraint describes the maximum number of processes that a manufacturer prioritizes for improvements. The next step of the quality improvement process is the selection of improvement actions. For every prioritized process k, the decision-maker finds a set of possible actions according to the domain knowledge.

In semiconductor manufacturing at Hitachi ABB, transistor chip production goes through 200 processes in a controlled environment. The raw material for this production is called the "'wafer", which is a thin slice of silicon, and 25 wafers generate an entity for the production. After the production of the wafer, it is cut into 62 chips for the final product. Hitachi ABB has 1197 production observations (M), and each production batch has $N = 3614$ production parameters from $K = 200$ different processes. After the production phase, quality testing is done for all the batches as a final step, and yield is calculated as the ratio of passing to failing chips. As the processes are complicated and numerous, it is difficult to find out the cause of yield loss manually. 3164 production parameters of 200 processes interact with each other and other operations, forming a nonlinear relationship. Hitachi ABB uses SHAP value methods to decompose prediction into contributions from each predictor. Traditional statistical process control is not helpful in explaining the interaction effects efficiently. Therefore, following the two-step approach explained above, the mean absolute feature attribution (Φ) can be calculated for each production parameter. A process's importance can easily be determined based on the highest feature attribution. As mentioned in Senoner et al. (2022), two processes are selected using a decision model that ranks the production parameters and the associated processes

164   8 Games in Characteristic Form: Applications in OM

based on the mean absolute feature attribution. Further, the selection of improvement actions is done for the production parameters of these processes.

8.6 Aumann–Shapley Pricing[12]

In Sect. 8.2, we discussed the benefit of pooling the nonidentical products through the concept of *economics of scope*. Characteristic-form game is *subadditive* when $V(S \cup T) \le V(S) + V(T)$ where S and T are disjoint sub-coalitions. This concept is widely applicable in a joint cost setting as sometimes producing the products together is cheaper than producing them separately. The cost subadditivity exists when

$$C\left(\sum s\right) \le \sum C(s)$$

where C is the cost function and s is the vector of goods required to be produced. A joint cost situation also arises when production is a nonseparable function for two or more products. However, the allocation of this cost among the contributing products is challenging. This is particularly difficult for a service produced in bulk when the users are in large numbers (or "infinitesimal"). Billera et al. (1978) address the cost allocation problem using *nonatomic games*. Aumann and Shapley (1974) provide desiderata for nonatomic game theory, which is game theory with a continuum of players. Billera et al. (1978) argue that for a system where it is impossible for any single customer to consume the entire service due to its high initial cost, customers join together to consume the available service. This translates the scenario to cooperative games where coalitions are formed, and the challenge is to allocate the cost fairly. However, the problem complexity increases multifold due to the continuum of players, and the solution concepts like the core and the Shapley value (refer Chap. 7) defined for characteristic games with finite players are not applicable. Aumann and Shapley (1974) extend the Shapley value to nonatomic games, and the modified Shapley values are called the *Aumann–Shapley prices* in the cost allocation setting (Young, 1985a).

Aumann and Shapley (1974) formally define nonatomic games as a model for competitive situations in which infinitely many participants are present, and no individual participant has any appreciable influence on the game. We consider a continuum of players represented as unit interval of $I = [0, 1]$, with σ-algebra $\mathcal{F}$ of subsets of $[0, 1]$. Each subset in $\mathcal{F}$ is *measurable*.[13] The members of I are called players and σ-algebra $\mathcal{F}$ is the set of coalitions. A real-valued function v on $\mathcal{F}$ to the real line ($v : \mathcal{F} \mapsto \mathbb{R}$) is the value of each coalition in $\mathcal{F}$. A nonatomic game is represented as $\langle I, \mathcal{F}, v \rangle$.

If n is a vector of goods or services that can be consumed by a continuum of players $[0, 1]$, Aumann and Shapley (1974) define $v = f \circ \mu$. f is a continuously

---

[12] In this section, the OM actions are on the *process side* in the VCAP framework.

[13] We recommend (Royden, 1988) for understanding measure theory and functional analysis.

8.6 Aumann–Shapley Pricing

differentiable function of n variables, with $f(0) = 0$. $\mu = (\mu_1, \mu_2, \ldots, \mu_n)$ is n-dimensional vector of measures defined on σ-algebra $\mathcal{F}$ as $\mu : \mathcal{F} \mapsto [0, 1]^n$. Aumann and Shapley (1974) define the Shapley value ϕ of coalition $S \in \mathcal{F}$ in nonatomic game $\langle I, \mathcal{F}, v \rangle$ as given in Eq. 8.15.

$$\phi(f \circ \mu)(S) = \sum_{j=1}^{n} \mu_j(S) \int_0^1 \frac{\partial f(t\mu(I))}{\partial x_j} dt \qquad (8.15)$$

For joint cost setting, we consider a finite set of different goods or services denoted as $N = \{1, 2, \ldots, n\}$ that are produced or consumed. A real-valued function C defined on $\mathbb{R}_+^n$ is the minimum cost of producing or offering the joint good by the provider, with $C(0) = 0$, and is continuously differentiable. Also, C is nondecreasing in demand vector $\alpha = (\alpha_1, \ldots, \alpha_n)$, which means $(\alpha \leq \alpha' \Rightarrow C(\alpha) \leq C(\alpha'))$ and $C(\alpha)$ is the joint cost of producing demand α. These n goods and services are consumed by a continuum of players, and the joint cost needs to be allocated to the consumers in a fair manner. This cost allocation leads to cost-based pricing mechanism $\mathbf{P} = (P_1, P_2, \ldots, P_n)$ where P_1 is the unit price of good (or service) 1 that fairly reflects the cost contribution of good 1 in the joint cost. Furthermore, the pricing mechanism $\mathbf{P}$ should satisfy the efficiency axiom of the Shapley value for budget balancedness, which means $P_1\alpha_1 + \cdots + P_n\alpha_m = C(\alpha)$ for demand vector α.

The ideas from the Shapley value of nonatomic games (Eq. 8.15) can be used to compute the desired cost-based pricing mechanism. Akin to the axioms to derive the Shapley value in Sect. 7.3.2, Billera and Heath (1982) and Mirman and Tauman (1982) use axiomatic approach to compute the desired pricing mechanism, and the unique price mechanism that satisfies the axioms is called *Aumann-Shapley prices*. The axioms considered are

Axiom I: Cost Sharing: For any output vector α, $\langle \mathbf{P}, \alpha \rangle = C(\alpha)$.
Axiom II: Rescaling: If $C'(\alpha) = C(\lambda\alpha)$, then $\mathbf{P}(C') = \lambda\mathbf{P}(C)$.
Axiom III: Additivity: If, $C = C_1 + C_2$, then $\mathbf{P} = \mathbf{P}_1 + \mathbf{P}_2$.
Axiom IV: Positivity: If cost function is increasing, then $\mathbf{P} \geq 0$.
Axiom V: Consistency: If all goods are same $P_1 = P_2 = \cdots = P_m$.

Theorem 8.4 (Samet & Tauman, 1982) *The Aumann–Shapley pricing is the unique cost-allocation-based pricing mechanism that satisfies Axioms I–V for all differentiable cost functions C with $C(0) = 0$. The pricing mechanism is given by*

$$P_i = \int_0^1 \frac{\partial C(t\alpha_1, t\alpha_2, \ldots, t\alpha_n)}{\partial \alpha_i} dt \qquad i = 1, \ldots, n \qquad \blacktriangleleft$$

To compute Aumann–Shapley price of a good, the marginal contribution of each infinitesimal unit of the good is calculated and the average is taken over each contribution. Additionally, a perfectly sampled coalition structure is calculated using the form $(t\alpha_1, t\alpha_2, \ldots, t\alpha_n)$, where t is a real number with $0 \le t \le 1$. The term $\frac{\partial C(t\alpha_1, t\alpha_2, \ldots, t\alpha_n)}{\partial \alpha_i}$ is the partial derivative of joint cost function C with respect to good i denoting the marginal cost contribution of the good. Examples 8.1 and 8.2 illustrate computation of Aumann–Shapley pricing.

Example 8.1 Consider a joint production process for producing two goods denoted as 1 and 2. The cost function to produce x_1 amount of good 1 and x_2 amount of good 2 is given by

$$F(x_1, x_2) = x_1^{\beta_1} + x_2^{\beta_2} + Ax_1x_2$$

where β_1, β_2, and A are positive constant parameters $(\beta_1, \beta_2, A > 0)$. To find the marginal cost contribution of each good is calculated by finding the partial derivative of the cost function, as shown below:

$$\frac{\partial F}{\partial x_1} = \beta_1 x_1^{\beta_1-1} + 0 + Ax_2$$

Similarly,

$$\frac{\partial F}{\partial x_2} = \beta_1 x_2^{\beta_2-1} + 0 + Ax_1$$

Now, to calculate prices for each product i for a production vector $\alpha = (\alpha_1, \alpha_2)$, we can use the A-S pricing formula in the following manner:

$$P_i = \int_0^1 \frac{\partial F(t\alpha_1, t\alpha_2)}{\partial x_i} dt$$

Applying the above pricing formula to the marginal cost contributions,

$$P_1 = \int_0^1 \left\{ \beta_1 (t\alpha_1)^{\beta_1-1} + A(t\alpha_2) \right\} dt$$

$$= \beta_1 \int_0^1 \alpha_1^{\beta_1-1} t^{\beta_1-1} dt + A \int_0^1 \alpha_2 t \, dt$$

$$= \beta_1 \alpha_1^{\beta_1-1} \cdot \frac{t^{\beta_1}}{\beta_1} \Big|_0^1 + A\alpha_2 \cdot \frac{t^2}{2} \Big|_0^1$$

$$= \alpha_1^{\beta_1-1} + \frac{A\alpha_2}{2}$$

Similarly, the price for product 2 can be found as $P_2 = \alpha_2^{\beta_2-1} + \frac{A\alpha_1}{2}$. The prices cover the cost and are fair. The efficiency axiom for pricing, discussed above, can be checked using the following approach to verify the cost-covering condition.

8.6 Aumann–Shapley Pricing

$$\langle \boldsymbol{\alpha}, \mathbf{P} \rangle = \alpha_1 P_1 + \alpha_2 P_2$$
$$= \alpha_1 \times \alpha_1^{\beta_1 - 1} + \frac{A\alpha_1\alpha_2}{2} + \alpha_2 \times \alpha_2^{\beta_2 - 1} + \frac{A\alpha_2\alpha_1}{2}$$
$$= \alpha_1^{\beta_1} + \alpha_2^{\beta_2} + A\alpha_1\alpha_2$$

◁

Example 8.2 (*Telephone Billing Game*) The telephone billing game was introduced by Billera et al. (1978) for Cornell University telephone system. Long-distance telephone service was divided into the following configurations in the United States:

Direct Distance Calling (DDD): Calls are charged on the duration of the call and the distance of the destination in a straightforward manner.

Foreign Exchange (FX): The charges include the monthly fixed fee for the line rent with the addition of the regular cost of the telephone at the foreign exchange.

Wide Area Telephone Service (WATS): United States was divided into five bands where Band $i + 1$ contains all the bands below it, i.e., b and 5 contained bands $i = 1, 2, 3, 4$. The charges involved a high initial fee and an incremental charge for exceeding the free limit.

Most of the cost components for the telephone service were fixed in nature, and it was difficult to divide among the class straightforwardly. As mentioned in the axioms of cost-based pricing mechanisms, the costs should be fully allocated among the service users, i.e., the prices must cover the expenses, and the prices should be "fair"—two calls made to the same destination during the same period should be charged same regardless of the person or the office.

Billera et al. (1978) use Aumann–Shapley prices games to compute the cost-based pricing mechanism. The calls can be segregated based on the locations (destinations), time of day, and type of day, as illustrated in Fig. 8.12. If two types of days, weekday and weekend, are considered, and k different locations to which the calls are placed with 24 hour bands, then we have $n = 24 \times k \times 2$ different *types* of calls. The set

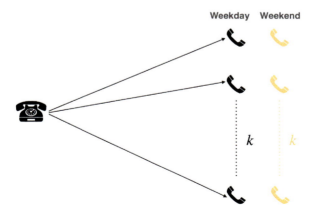

Fig. 8.12 Illustration of Telephone Call Segregation System

of players I contains sub-intervals $I_j = [j - 1, j)$ representing the calls of type $j \in \{1, \ldots, n\}$. μ_j is the measure defined on the measurable space $(I, \mathcal{F})$, which gives the number of minutes of the call for sub-coalition S denoted as $\mu_j(S)$. For instance, $\mu_{23}(S)$ measures the total number of minutes of the calls in S that are placed on WATS band 3 for the interval 2 to 3 am on a weekday. $\mu = (\mu_1, \mu_2, \ldots, \mu_n)$ is the vector of measures of call duration.

The pricing problem starts with the optimization routine where the least cost of providing the service is calculated (Lampell, 1977). Given the load on the system denoted by $X = (x_1, x_2, \ldots, x_n)$, a least cost function for S, $f(x_1, x_2, \ldots, x_n)$ is derived from the optimization module. Telephone billing game is nonatomic game $\langle I, \mathcal{F}, v \rangle$ with $v \equiv f \circ \mu = f(\mu(S))$. The Shapley value computed using the expression in Theorem 8.4 gives Aumann–Shapley price $p_j = \int_0^1 f_j(t(\mu(I))dt$ for each call type j, where $f_j = \partial f / \partial x_j$. ◁

Aumann–Shapley pricing has wide applications in pricing public utilities. The consistency axiom recommends same price for the goods that incur the same cost. McLean and Sharkey (1998) provide a method to discriminate prices among the same goods even if the cost contribution is equal. This modified approach is useful in regulatory pricing, where commercial users are charged a higher fee than contemporary domestic ones. Calvo and Santos (2001) and Sprumont (2005) propose mixed and discretized versions of Aumann–Shapley prices, respectively.

Problems

Problem 8.1 Consider service system cooperative game $\langle N, V \rangle$ with $N = \{1, 2, 3\}$, $\alpha = 1$, $\{\lambda_1, \lambda_2, \lambda_3\} = \{1, 1, 5\}$, and $\{\xi_1, \xi_2, \xi_3\} = \{10, 10, 10\}$. Find the Shapley value of the game. Check whether it is in the core. If not, compute the Shapley value for the reduced game $\langle N, W \rangle$. Interpret your results. ◀

Problem 8.2 In the recycling network game (Sect. 8.3), consider $N = 5$, $\bar{c}(n) = 0.5(1 - 1.2^{1-n})$. Simulate the dynamically stable coalitions for the following values: level of heterogeneity $x = [0, 1]$, and fixed recycling cost $K = [0, 50000]$.

1. What is the cutoff value of K after which the grand coalition is dynamically stable irrespective of x?
2. What are the respective preferable ranges of x and K for the network $\{2, 3\}$ and $\{1, 4\}$ to be dynamically stable? ◀

Problem 8.3 In the figure, Player 1 controls links $\{(s, a), (a, e), (e, t)\}$. Player 2 controls $\{(s, b), (s, c), (b, f), (c, f), (f, t)\}$. The rest of the links are controlled by Player 3. Using Shapley value, compute the load distributions for the three players.

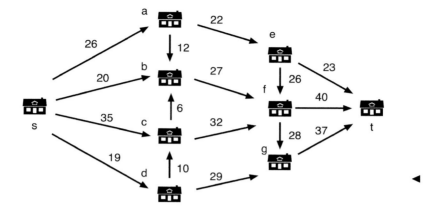

References

Anily, S., & Haviv, M. (2010). Cooperation in service systems. *Operations Research, 58*(3), 660–673.

Atasu, A., Lifset, R., Linnell, J., Perry, J., Sundberg, V., Mayers, C. K., Dempsey, M., Van Wassenhove, L. N., van Rossem, C., & Gregory, J., et al. (2010). Individual producer responsibility: A review of practical approaches to implementing individual producer responsibility for the WEEE directive.

Aumann, R. J., & Shapley, L. S. (1974). *Values of non-atomic games*. Princeton University Press.

Bendel, D., & Haviv, M. (2018). Cooperation and sharing costs in a tandem queueing network. *European Journal of Operational Research, 271*(3), 926–933.

Billera, L. J., & Heath, D. C. (1982). Allocation of shared costs: A set of axioms yielding a unique procedure. *Mathematics of Operations Research, 7*(1), 32–39.

Billera, L. J., Heath, D. C., & Raanan, J. (1978). Internal telephone billing rates-a novel application of non-atomic game theory. *Operations Research, 26*(6), 956–965.

Calvo, E., & Santos, J. C. (2001). Prices in mixed cost allocation problems. *Games and Economic Behavior, 37*(2), 243–258.

Chiou, C.-C. (2008). Transshipment problems in supply chain systems: review and extensions. *Supply Chain,* 427–448.

Chwe, M. S.-Y. (1994). Farsighted coalitional stability. *Journal of Economic theory, 63*(2), 299–325.

Ergün, S., Usta, P., Alparslan Gök, S. Z., & Weber, G. W. (2021). A game theoretical approach to emergency logistics planning in natural disasters. *Annals of Operations Research,* 1–14.

Gopalakrishnan, S., Granot, D., Granot, F., Sošić, G., & Cui, H. (2021). Incentives and emission responsibility allocation in supply chains. *Management Science, 67*(7), 4172–4190.

Gui, L., Atasu, A., Ergun, Ö., & Toktay, L. B. (2015). Efficient implementation of collective extended producer responsibility legislation. *Management Science,* Article in Press.

Guo, P., Leng, M., & Wang, Y. (2013). A fair staff allocation rule for the capacity pooling of multiple call centers. *Operations Research Letters, 41*(5), 490–493.

Hafezalkotob, A., & Makui, A. (2015). Cooperative maximum-flow problem under uncertainty in logistic networks. *Applied Mathematics and Computation, 250,* 593–604.

Karsten, F., Slikker, M., & Van Houtum, G.-J. (2015). Resource pooling and cost allocation among independent service providers. *Operations Research, 63*(2), 476–488.

Kemahlıoğlu-Ziya, E., & Bartholdi, J. J., III. (2011). Centralizing inventory in supply chains by using Shapley value to allocate the profits. *Manufacturing & Service Operations Management, 13*(2), 146–162.

Lampell, D. (1977). On the selection of numbers of servers for the n server-type problem (thesis). Technical report, Cornell University Operations Research and Industrial Engineering.

Lundberg, S. M., & Lee, S.-I. (2017). A unified approach to interpreting model predictions. *Advances in Neural Information Processing Systems, 30.*

McLean, R. P., & Sharkey, W. W. (1998). Weighted Aumann-Shapley pricing. *International Journal of Game Theory, 27,* 511–523.

Mirman, L. J., & Tauman, Y. (1982). Demand compatible equitable cost sharing prices. *Mathematics of Operations Research, 7*(1), 40–56.

Panzar, J. C., & Willig, R. D. (1981). Economies of scope. *The American Economic Review, 71*(2), 268–272.

Reyes, P. M. (2005). Logistics networks: A game theory application for solving the transshipment problem. *Applied Mathematics and Computation, 168*(2), 1419–1431.

Royden, H. L. (1988). *Real analysis.* Prentice Hall of India.

Rudramoorthi, T., & Amit, R. K. (2023). Repositioning game for ambulance services. *Transportation Research Record, 2677*(6), 113–128.

Russell, S. J., & Norvig, P. (2020). *Artificial intelligence : A modern approach* (4th ed.). Illustrationen, SE - XVII, 1115 Seiten .

Samet, D., & Tauman, Y. (1982). The determination of marginal cost prices under a set of axioms. *Econometrica: Journal of the Econometric Society,* 895–909.

Schmenner, R. W., & Swink, M. L. (1998). On theory in operations management. *Journal of Operations Management, 17*(1), 97–113.

Senoner, J., Netland, T., & Feuerriegel, S. (2022). Using explainable artificial intelligence to improve process quality: Evidence from semiconductor manufacturing. *Management Science, 68*(8), 5704–5723.

Sprumont, Y. (2005). On the discrete version of the aumann-shapley cost-sharing method. *Econometrica, 73*(5), 1693–1712.

Taguchi, G. (1986). *Introduction to quality engineering: Designing quality into products and processes.*

Tian, F., Šošić, G., & Debo, L. (2020). Stable recycling networks under the extended producer responsibility. *European Journal of Operational Research, 287*(3), 989–1002.

Young, H. P. (1985a). *Cost allocation: Methods, principles, applications.* North Holland Publishing Co.

Young, H. P. (1985). Monotonic solutions of cooperative games. *International Journal of Game Theory, 14*(2), 65–72.

Chapter 9
Mechanism Design and Auctions

In Chap. 2, we discussed the primitives of game theory. Apart from the agents and their action sets, the primitives include the rules (or protocols) that enable interaction among the players and information about the environment in which a game is embedded. Given the rules and the environment, the joint action of players leads to an outcome, and each player has a preference over the outcomes. Different solution concepts for noncooperative games prescribe equilibrium outcomes. Communication among the players is modeled in noncooperative games, and the rules include communication protocols among the agents.

In many settings, information about the environment is not publicly observable. An example is auctioning a good for which each buyer has a private valuation, which is not known to the seller. However, the seller has a desirable goal to sell it to the buyer with the highest valuation. Is it possible for the seller to design the auction as a game where the interaction among the strategic buyers reveals the buyer with the highest valuation? *Mechanism design* deals with designing games by defining the rules, the action sets for the agents, and the appropriate equilibrium concept such that the designer's desired outcomes are the subset of the equilibrium outcomes of the game.

9.1 Examples

We begin this chapter with examples to improve our understanding.

9.1.1 Procurement Problem

Some of these examples are motivated by examples from Kreps (1990).

© The Author(s), under exclusive license to Springer Nature Singapore Pte Ltd. 2024
R. K. Amit, *Game Theory with Applications in Operations Management*, Springer Texts in Business and Economics, https://doi.org/10.1007/978-981-99-4833-8_9

Example 9.1 (*Vaccine Procurement*) Consider a procurement setting where a government has to procure 100 million vaccine doses. Firm A and Firm B are the only two manufacturers of the vaccine. The cost of producing a single dose is not publicly observable and is private information for each firm. However, it is common knowledge that the cost per dose in dollars can be either *one* or *two*, and each cost is equally likely. The setting is illustrated in Fig. 9.1.

Government has to procure 100 million vaccine doses at lowest possible cost

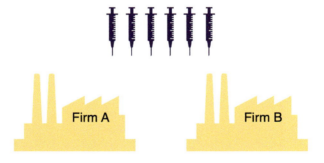

$C_i = \{1, 2\}$ in dollars. Each cost is equally likely.
$i = A, B$
Cost information is private information

Fig. 9.1 Vaccine procurement

The government's desired goal is to procure the vaccine doses at the lowest cost. In this example, the environment has four possible *states*—$(C_A = 1, C_B = 1)$, $(C_A = 1, C_B = 2)$, $(C_A = 2, C_B = 1)$, $(C_A = 2, C_B = 2)$, but the government is not aware which is the true state. The desired procurement function (a mapping from the states to the outcomes) of the government is

$$f(C_A = 1, C_B = 1) = (x_A = 50, x_B = 50)$$
$$f(C_A = 1, C_B = 2) = (x_A = 100, x_B = 0)$$
$$f(C_A = 2, C_B = 1) = (x_A = 0, x_B = 100)$$
$$f(C_A = 2, C_B = 2) = (x_A = 50, x_B = 50)$$

9.1 Examples

The procurement function indicates that the government desires to procure 100 million doses from the firm that has the lower cost. If the costs are equal, the government desires to procure 50 million doses from each firm. In the desired procurement function, x_i is the doses procured from firm i ($i = A, B$).

The government wishes to *implement* the desired procurement function by designing the procurement mechanism as a game with rules and action sets for each firm such that the interaction of the firms in the game reveals the true state as an equilibrium of the game. ◁

Example 9.2 (*Vaccine Procurement,* Naive Mechanism) For the vaccine procurement setting in Example 9.1, let the government use a mechanism, which we call *naive mechanism*, and is shown in Fig. 9.2.

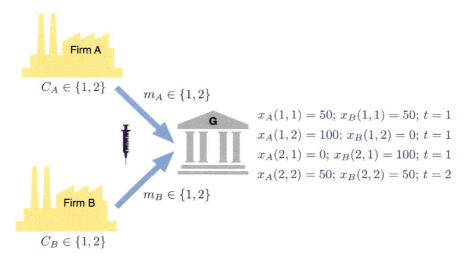

Fig. 9.2 Vaccine procurement—Naive mechanism

In this mechanism, the government asks them to reveal their costs by choosing an action $m_i \in \{1, 2\}$, which messages each firm's cost to the government. The government considers Bayesian-Nash equilibrium as the appropriate solution concept. The rules include simultaneous choice of actions and the procurement scheme based on the choice of actions. As shown in Fig. 9.2, the procurement scheme is—if they message the same cost, the order is split, and each firm is paid the messaged cost; and if their messages differ, the complete order is given to the firm with lower cost, and paid its messaged cost. For example, if $m_A = 1$ and $m_B = 2$, then the procurement scheme orders 100 million doses from firm A ($x_A(1, 2) = 100$; $x_B(1, 2) = 0$) at a price $t = 1$ dollar per dose.

Will the firms reveal their true costs as Bayesian-Nash equilibrium in Naive mechanism? If they reveal their true costs, the government can implement the desired procurement function in Bayesian-Nash equilibrium using Naive mechanism. Unfortunately, the answer is no.

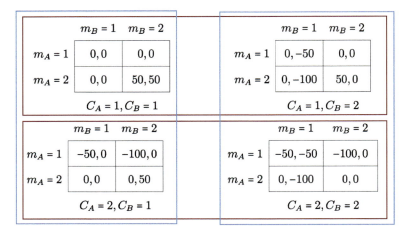

Fig. 9.3 Bayesian game of Naive procurement mechanism

The Bayesian game associated with Naive procurement mechanism is shown in Fig. 9.3. As mentioned in Example 9.1, the environment has four possible states, each with probability 0.25, and each state leads to a different bi-matrix game in Fig. 9.1. The payoffs are computed based on the choice of m_A and m_B, given their respective costs. For example, in the game with $C_A = 1, C_B = 1$, if both the firms message 2, then the order is divided, and each makes a profit of 50 million dollars. The red boxes are the information partition of Firm A, and the blue boxes are the information partition of Firm B. The top red box is the information partition of Firm A when $C_A = 1$.

Let us compute the equilibria of the game. In any state (or game), if a firm's cost is 2, then messaging cost as 2 strictly dominates messaging 1 for that firm; and, if a firm's cost is 1, then messaging cost as 2 weakly dominates messaging 1 for that firm. For example, in the top red information partition when $C_A = 1$, Firm A *conjectures* that Firm B messages strictly dominant strategy in the game with $C_A = 1, C_B = 2$ and messages weakly dominant strategy in the game with $C_A = 1, C_B = 1$, then the best response of Firm A is to message $m_A = 2$. This reasoning is also applicable for Firm B when $C_B = 1$. Also, we have mentioned that when $C_A = 2$, messaging $m_A = 2$ is strictly dominant, and likewise for Firm B when $C_B = 2$. This means, in Bayesian-Nash equilibrium of naive mechanism, each firm messages cost as 2, irrespective of its true cost. The messages are uninformative, and the government, still not knowing the true costs, cannot implement the desired procurement function.

Before designing the mechanism, the government's expected cost of procurement is $0.25 \times (100 + 100 + 100) + 0.25 \times 200 = 125$ million dollars. The expected procurement cost in naive mechanism is 200 million dollars. In naive mechanism, the incentives are not *compatible*[1] for the firms to reveal their true costs. Can the government *engineers*[2] the incentives for the firms to reveal their true costs?   ◁

Example 9.3 (*Vaccine Procurement,* Improved Mechanism 1) In Example 9.2, the government could not implement the desired procurement function using Naive mechanism, as the incentives are not compatible for the firms to reveal their true costs. The government uses an improved mechanism, which we call *Improved Mechanism 1* (or IM1), and is shown in Fig. 9.4. In IM1, when each firm messages cost as 1, the price per dose is increased to 2 dollars from 1 dollar in naive mechanism. IM1 provides an incentive to the firms to message the cost as 1, when the true cost is 1. The other rules and the solution concept are the same as in Naive mechanism.

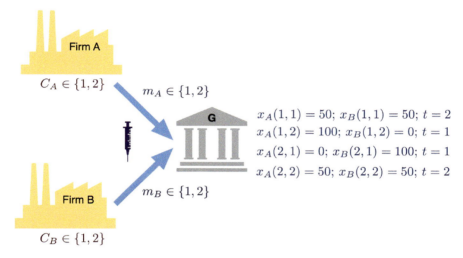

Fig. 9.4 Vaccine procurement—Improved Mechanism 1

Will the firms reveal their true costs as Bayesian-Nash equilibrium in IM1? The Bayesian game associated with IM1 is shown in Fig. 9.5. If a firm's cost is 2, then messaging cost as 2 is a weakly dominant strategy for that firm. Now, consider the top red information partition when $C_A = 1$, and if Firm A conjectures that Firm B messages $m_B = 2$, independent of its cost, then the best response of Firm A is to message $m_A = 2$. This reasoning is also applicable for Firm B when $C_B = 1$. This means, in this Bayesian-Nash equilibrium of IM1, each firm messages cost as 2, irrespective of its true cost. The messages are uninformative, and the government, still not knowing the true costs, cannot implement the desired procurement function.

Fortunately for the government, there is another Bayesian-Nash equilibrium. Consider the top red information partition again when $C_A = 1$, Firm A conjectures that

---

[1] *Incentive compatibility* is formulated as a constraint when we design mechanisms.
[2] Mechanism design is also known as *incentive engineering*.

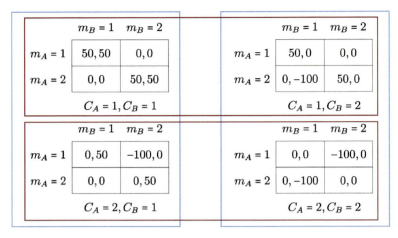

Fig. 9.5 Bayesian game of Improved Mechanism 1

Firm B always messages the true cost—$m_B = 1$ when $C_B = 1$, and $m_B = 2$ when $C_B = 2$, then messaging $m_A = 1$ is one of the best response of Firm A.[3] This reasoning is also applicable for Firm B when $C_B = 1$. We also know that if a firm's cost is 2, then messaging cost as 2 is a weakly dominant strategy for that firm. This implies that, in this Bayesian-Nash equilibrium, each firm is messaging its true cost. In IM1, there exists a Bayesian-Nash equilibrium in which each firm reveals its true cost, and the government can implement the desired procurement function.

The mechanism IM1 induces two equilibria—in one, the firms reveal the truth, and the expected procurement cost to the government is 150 million dollars; while in the other, each firm always messages cost as 2, irrespective of its true cost, and the expected procurement cost to the government is 200 million dollars. The government is not sure which of the equilibrium is selected by the firms and is uncertain about the procurement costs. Can the government improve the incentives to ensure a *unique* truth-revealing equilibrium? ◁

Example 9.4 (*Vaccine Procurement*, Improved Mechanism 2) In Example 9.3, the government implements the desired procurement function in one of the Bayesian-Nash equilibria using IM1. However, there exists another equilibrium that is not truth revealing, and this makes the government uncertain about the procurement costs. The government hires a consultant to design a mechanism that implements the desired procurement function in a unique Bayesian-Nash equilibrium. The consultant *guesses* the incentives and proposes a new mechanism, which is named *Improved Mechanism 2* (or IM2). IM2 is similar to IM1, except it uses a different payment scheme. IM2 is shown in Fig. 9.6.

---

[3] In this literature, it is assumed that when an agent is indifferent between two strategies, the tie is often resolved in favor of the mechanism designer.

9.1 Examples

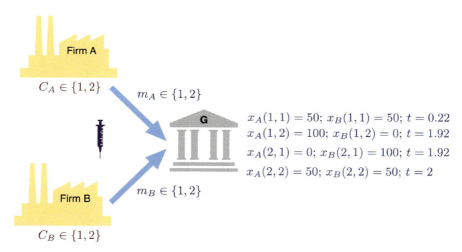

Fig. 9.6 Vaccine procurement—Improved Mechanism 2

	$m_B = 1$	$m_B = 2$
$m_A = 1$	−39, −39	92, 0
$m_A = 2$	0, 92	50, 50

$C_A = 1, C_B = 1$

	$m_B = 1$	$m_B = 2$
$m_A = 1$	−39, −89	92, 0
$m_A = 2$	0, −8	50, 0

$C_A = 1, C_B = 2$

	$m_B = 1$	$m_B = 2$
$m_A = 1$	−89, −39	−8, 0
$m_A = 2$	0, 92	0, 50

$C_A = 2, C_B = 1$

	$m_B = 1$	$m_B = 2$
$m_A = 1$	−89, −89	−8, 0
$m_A = 2$	0, −8	0, 0

$C_A = 2, C_B = 2$

Fig. 9.7 Bayesian game of Improved Mechanism 2

The Bayesian game associated with IM2 is shown in Fig. 9.7. If a firm's cost is 2, then messaging cost as 2 is a strictly dominant strategy for that firm. Now, consider the top red information partition when $C_A = 1$, and Firm A conjectures that Firm B messages the true cost $m_B = 1$ when $C_B = 1$, then messaging $m_A = 1$ is the best response of Firm A ($m_A = 1$ gives the expected payoff of 26.5 million dollars, while $m_A = 2$ gives the expected payoff of 25 million dollars). Using the same reasoning, in the left blue information partition of Firm B when $C_B = 1$, Firm B chooses $m_B = 1$. Hence, in the Bayesian game associated with IM2, there is a *unique* Bayesian-Nash equilibrium in which the firms reveal their true costs, with the expected procurement cost of 151.5 million dollars to the government. This is 1.5 million dollars higher than the expected procurement cost in the truth-revealing Bayesian-Nash equilibrium in IM1. This is the extra incentive for the unique equilibrium in IM2.

In IM1 and IM2, the government implements the desired procurement function in Bayesian-Nash equilibrium, but the implementation is conditional on truth revealing by a firm based on its *conjecture* of truth revealing by the other firm. To overcome the conjectural dependence, the government wishes to implement the desired procurement function in *dominant strategy equilibrium*—truth revealing is a dominant strategy for each firm, independent of the strategy of the other firm. ◁

Example 9.5 (*Vaccine Procurement,* Improved Mechanism 3) In Example 9.4, with the help of the consultant, the government implements the desired procurement function in unique Bayesian-Nash equilibrium using IM2. The government asks the consultant to improve the mechanism to implement the desired procurement function in dominant strategy equilibrium. The consultant again guesses the incentives and proposes a new mechanism, which we name *Improved Mechanism 3* (or IM3). IM3 is similar to the previous mechanisms, except for modification in the payment scheme, as shown in Fig. 9.8.

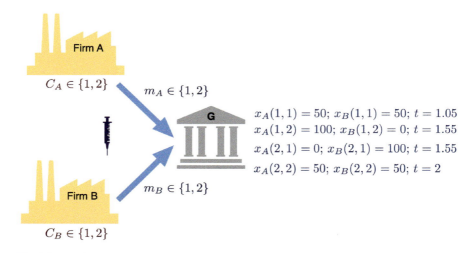

Fig. 9.8 Vaccine procurement—Improved Mechanism 3

The Bayesian game associated with IM2 is shown in Fig. 9.9.

9.1 Examples 179

	$m_B = 1$	$m_B = 2$
$m_A = 1$	2.5, 2.5	55, 0
$m_A = 2$	0, 55	50, 50

$$C_A = 1, C_B = 1$$

	$m_B = 1$	$m_B = 2$
$m_A = 1$	2.5, -47.5	55, 0
$m_A = 2$	0, -45	50, 0

$$C_A = 1, C_B = 2$$

	$m_B = 1$	$m_B = 2$
$m_A = 1$	-47.5, 2.5	-45, 0
$m_A = 2$	0, 55	0, 50

$$C_A = 2, C_B = 1$$

	$m_B = 1$	$m_B = 2$
$m_A = 1$	-47.5, -47.5	-45, 0
$m_A = 2$	0, -45	0, 0

$$C_A = 2, C_B = 2$$

Fig. 9.9 Bayesian game of Improved Mechanism 3

In the Bayesian game, in each state, messaging the true cost to the government is a *strictly dominant strategy*, independent of the choice of strategy by the other firm. In other words, the government implements the desired procurement function in dominant strategy equilibrium with incentives that are compatible with truth revealing. The expected procurement cost is 153.75 million dollars, which can be further reduced until the incentives are right for messaging the true cost as a dominant strategy for each firm. ◁

9.1.2 Mechanism Design: Comments

The theory of mechanism design has its origin in the debates of Hayek-Lange-Lerner on the supremacy between free markets and market socialism. Markets work by mobilizing the dispersed information through the price system, and prices serve as incentive devices in coordinating that dispersed information (Hayek, 1937, 1945). However, the critics of Hayek, Oskar Lange (Lange & Taylor, 1938) and Lerner (1944) popularized the idea of market socialism where a central economic planner could implement any allocation achieved by the competitive markets; and, moreover, could improve upon the allocations made by the competitive markets in the cases where the markets fail due to various imperfections. Their ideas motivated the research on mechanism design. Oskar Lange, in his paper "The computer and the market" (Lange, 1965), provided the early ideas for modern business platforms like Uber and connected them to the theory of mechanism design (Fig. 9.10). We will discuss the connection in Example 9.6 on contract nets.

THE COMPUTER AND THE MARKET

OSKAR LANGE

I

Not quite 30 years ago I published an essay *On the Economic Theory of Socialism*.† Pareto and Barone had shown that the conditions of economic equilibrium in a socialist economy could be expressed by a system of simultaneous equations. The prices resulting from these equations furnish a basis for rational economic accounting under socialism (only the static equilibrium aspect of the accounting problem was under consideration at that time). At a later date Hayek and Robbins maintained that the Pareto–Barone equations were of no practical consequence. The solution of a system of thousands or more simultaneous equations was in practice impossible and, consequently, the practical problem of economic accounting under socialism remained unsolvable.

In my essay I refuted the Hayek-Robbins argument by showing how a market mechanism could be established in a socialist economy which would lead to the solution of the simultaneous equations by means of an empirical procedure of trial and error. Starting with an arbitrary set of prices, the price is raised whenever demand exceeds supply and lowered whenever the opposite is the case. Through such a process of *tâtonnements*, first described by Walras, the final equilibrium prices are gradually reached. These are the prices satisfying the system of simultaneous equations. It was assumed without question that the *tâtonnement* process in fact converges to the system of equilibrium prices.

Were I to rewrite my essay today my task would be much simpler. My answer to Hayek and Robbins would be: so what's the trouble? Let us put the simultaneous equations on an electronic computer and we shall obtain the solution in less than a second. The market process with its cumbersome *tâtonnements* appears old-fashioned. Indeed, it may be considered as a computing device of the pre-electronic age.

Fig. 9.10 Lange's early ideas for platforms

Hurwicz (1960, 1972) point out that the mechanisms are the economic institutions by which the economic activity is to be coordinated, and consider there may be many possible mechanisms that can achieve the desired goals. Can we design an optimal mechanism that can achieve the desired goals? Leonid Hurwicz formalized the theory of mechanism design (Hurwicz, 1960, 1972), and for this seminal contribution, he shared the 2007 Nobel Memorial Prize in Economic Sciences with Eric Maskin and Roger Myerson (Nobel Prize in Economic Sciences, 2007).

In tandem with Hurwicz (1960, 1972), we have seen in Examples 9.3–9.5 that multiple mechanisms can implement the desired procurement function. Let us understand the wisdom from the examples in the context of the theory of mechanism design:

(a) In Example 9.1, the government wants to procure 100 million vaccine doses. Firm A and Firm B are the only two manufacturers of the vaccine. The cost of producing a single dose is not publicly observable and is private information for each firm. The government has a desired procurement function and wants to implement the desired procurement function using mechanisms so that the true production cost is revealed.

(b) In Example 9.2, the government could not implement the desired procurement function in Bayesian-Nash equilibrium using Naive mechanism, as the incentives are not compatible with truthfully revealing the production costs.

(c) In Example 9.3, the government modified the payment scheme and used IM1 to implement the desired procurement function. It was shown that IM1 induces Bayesian-Nash equilibrium that implements the desired procurement function. However, it also induces another Bayesian-Nash equilibrium that is not truth

9.1 Examples 181

revealing. In the mechanism design literature, this situation is called *weak implementation*, as the mechanism may have multiple equilibria, and only some of them implement the desired goals.

(d) In Example 9.4, the government used IM2, which implements the desired procurement function in a unique Bayesian-Nash equilibrium, but the implementation is conditional on truth revealing by a firm based on its conjecture of truth revealing by the other firm.

(e) In Example 9.5, the government revised the payment scheme in IM3, which implements the desired procurement function in dominant strategy equilibrium. In IM3, truthfully revealing its true cost is a dominant strategy, independent of the truthful revelation of the cost by the other firm.

As obvious from the examples, implementation in dominant strategy equilibrium is the *most preferred implementation* for the mechanism designer as it mitigates the risk of the conjectural dependence of truth revealing by the other players. However, there is a caveat with implementation in dominant strategy equilibrium. The impossibility result of the *Gibbard–Satterthwaite theorem* (Gibbard, 1973; Satterthwaite, 1975) restricts the choice of the procurement (or social choice or allocation) functions[4] that are implementable in dominant strategy equilibrium. The Gibbard–Satterthwaite theorem states that if we allow a sufficiently broad domain of individual preferences, then only *dictatorial social choice functions*[5] can be implemented in dominant strategies. For implementation in dominant strategies, the domain is restricted to *quasilinear preferences*—an agent's preference depends only on the outcomes of the mechanism and her own payment, and she does not care about the payments to the other agents. For example, in single-unit auctions, bidder i's quasilinear preferences can be represented as $v_i - p_i$, where v_i is the utility to the bidder if she wins the auction and p_i is the payment made by her.

We recommend Moore (1992), Mas-Colell et al. (1995), Palfrey (2002), and Shoham and Leyton-Brown (2009) for further reading on mechanism design.

9.1.3 More Examples

Mechanism design has numerous applications, as shown in Fig. 9.11. We discuss some of them in the following examples.

---

[4] A social choice function is a mapping from individual preferences to the set of outcomes. The desirable procurement function in the earlier examples is a type of social choice function. In Examples 9.2–9.5, it can be observed that when the state changes, the firms' preferences change, and the desired procurement function maps the possible states to the outcomes.

[5] A social choice function is dictatorial if there is an agent whose most preferred choice is always the output of the social choice function.

182   9 Mechanism Design and Auctions

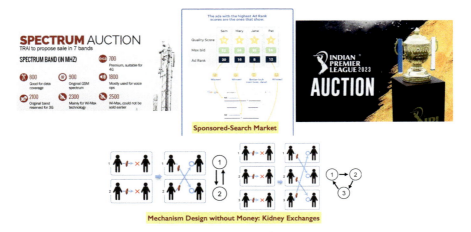

Fig. 9.11 Applications of mechanism design

Example 9.6 (*Contract Net*) Smith (1981) introduce contract net protocol to model the contract-making process by the companies. The contract net protocol has four stages, as shown in Fig. 9.12.

In the *task recognition* stage, the principal recognizes a task (painting a house in Fig. 9.12) that needs support from the agents. In the *task announcement* stage, the principal announces the task to the agents. This announcement should include sufficient information about the task, such as deadlines and quality of service, to enable the agents to decide whether to participate and bid. In the bidding stage, the agents bid to execute the task based on their capabilities. In the *contract award* stage, the principal, based on the submitted bids, awards the contract to the most appropriate agent. The successful agent, while executing the task, may recognize new tasks whose execution follows the same protocol. This forms a *net* of contracts.

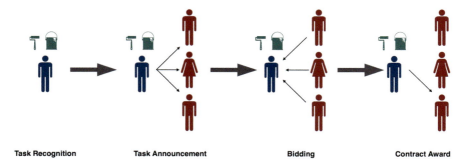

Fig. 9.12 Contract net protocol (Adapted from Russell & Norvig, 2020)

9.1 Examples

As evident, the contract net protocol is a mechanism design problem and is used in modern business platforms like Uber. *Urban Company* is one such business platform in India to hire individuals for home painting. We recommend Davis and Smith (1983), Sandholm (1993, 1998) for further reading on contract nets. ◁

Example 9.7 (*The Sage's Judgment*[6]) In a *Jataka* tale,[7] two women claimed to be the true mother of a child. They approached a sage to resolve the conflict. The sage drew a line and laid the child in the middle of the line. He then asked one woman to seize the child by the hands and the other woman by the feet, as shown in Fig. 9.13. He announced to them, "Lay hold of it and pull; the child is hers who can pull it over." They both pulled, and the child, being pained while pulling, cried loudly. Seeing the child crying, the true mother lets the child go. Based on their behavior, the sage identified the true mother.

We can represent the story in the mechanism design framework. Let the set of outcomes be $\mathcal{A} = \{a, b, c\}$, where a means that the child is given to woman 1, b means that the child is given to woman 2, and c is the child crying. The set of states $\Theta = \{\theta_1, \theta_2\}$, where θ_1 is the state when woman 1 is the true mother, and θ_2 is the state when woman 2 is the true mother. The sage can assume that when the state changes, the preferences over the set of outcomes change. For example, in state θ_1, woman 2 prefers outcome b over outcome c over outcome a. The preferences are shown below:

$$\theta_1: \quad a \succ_1 b \succ_1 c \quad b \succ_2 c \succ_2 a$$
$$\theta_2: \quad a \succ_1 c \succ_1 b \quad b \succ_2 a \succ_2 c$$

The desired social choice function of the sage is $f(\theta_1) = a$ and $f(\theta_2) = b$. The sage is unaware of the true state and designs the discussed mechanism to know "who is the true mother?". ◁

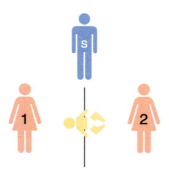

Fig. 9.13 The Sage's judgment

---

[6] This is similar to the biblical story of the judgment of King Solomon. However, the mechanism used in the Jataka story is different. Moore (1992) discusses King Solomon's judgment as a mechanism design problem in detail.

[7] Source: *The Jataka*, Vol. VI, by E. B. Cowell and W. H. D. Rouse (1907).

Example 9.8 (*Sealed-bid Auctions*) Since ancient times, auctions have been used to determine the value of something through competitive bidding.[8] Cassady (2021) provides numerous examples of how auctions are used in different settings and in different countries.

We discussed first-price sealed-bid auction (FPSB auction) in Example 3.4 and second-price sealed-bid auction (SPSB auction) in Example 3.10. It was shown that, in FPSB auction with uniformly distributed private valuations between 0 and 1, the bidders bid half of their valuations in Bayesian-Nash equilibrium. FPSB auction with two bidders is shown in Fig. 9.14.[9]

We also showed that in SPSB auction, bidding your private valuation is a dominant strategy, as shown in Fig. 9.15. The sealed-bid auctions are mechanism design problems and have similarities with the vaccine procurement problem.[10] In FPSB and SPSB auctions, the good is auctioned to the bidder with the highest valuation; however, FPSB and SPSB auctions use different payment functions. Furthermore, SPSB auction is truth revealing in dominant strategies, which is the most preferred implementation, as discussed in Sect. 9.1.2; while FPSB auction is not truth revealing even in Bayesian-Nash equilibrium. We discuss auction design in Sect. 9.3. ◁

Fig. 9.14 First-price sealed-bid auction

---

[8] In India, *Jain auctions* are used in Jainism "to serve both religious ends (pragmatic and karmic) and social ends (establishing and protecting social status among bidders)" (Kelting, 2009, p. 285).

[9] In the figure, the seller is auctioning IBM Simon, the first smartphone released in 1994.

[10] Auctions for procurement are called *reverse auctions* with one buyer and multiple sellers; while the auctions discussed in this example are *forward auctions* with one seller and multiple buyers.

9.2 Mechanism Design: Formalism

Fig. 9.15 Second-price sealed-bid auction

9.2 Mechanism Design: Formalism

In this section, we discuss the formal structure of the mechanism design problem. For notational brevity, we consider two agents $\mathcal{N} = \{1, 2\}$ and finite alternatives $\mathcal{A} = \{a_1, \ldots, a_k\}$. The discussion can be easily extended to n agents. Θ be the set of states of the environment with θ be a typical state. In Examples 9.1–9.5, there are four possible states. A *social choice rule* F is a mapping $F : \Theta \mapsto 2^{\mathcal{A}} \setminus \{\emptyset\}$. A social choice rule, if single-valued, is called a *social choice function*. The principal wishes to *implement* a social choice rule: in each state, the principal wishes to realize some desired outcomes. Unfortunately, in incomplete information settings, the principal does not know the true state. Hence, for successfully implementing a desired social choice rule under incomplete information, we require a *mechanism*.

A mechanism $\mathcal{M} = ((M_1, M_2), g)$ describes a *message* or *strategy set* M_i for agent i and an *outcome function* $g : M_1 \times M_2 \mapsto \mathcal{A}$. In Examples 9.1–9.5, $M_i = \{1, 2\}$ for $i = A, B$. An agent can conjecture what other agent's message might be, and that is the role for different *solution concepts* $\mathcal{C}$ (Nash equilibrium, Bayesian-Nash equilibrium, subgame perfect Nash equilibrium,[11] dominant strategies, etc.) that are employed. The mechanism induces a different game in each state as the agents' preferences over outcomes may change with states. Assume $(\mathcal{M}, \theta)$ is the game induced by the mechanism $\mathcal{M}$ in the state θ. $\mathcal{C}(\mathcal{M}, \theta)$ gives the set of message profiles that are recommended by $\mathcal{C}$ in the game $(\mathcal{M}, \theta)$. The corresponding set of outcomes will be denoted as $g(\mathcal{C}(\mathcal{M}, \theta))$. For example, if Bayesian-Nash equilibrium

---

[11] The principal can use games in extensive form as mechanisms (Moore & Repullo, 1988).

is the solution concept, $\mathcal{C}(\mathcal{M}, \theta)$ gives us the set of Bayesian-Nash equilibria of the game induced by $(\mathcal{M}, \theta)$; and $g(\mathcal{C}(\mathcal{M}, \theta))$ gives the corresponding set of Bayesian-Nash equilibrium outcomes.

Definition 9.1 (*Implementation*) A social choice rule F is $\mathcal{C}$-**implementable** if there exists a mechanism $\mathcal{M}$ such that for every $\theta \in \Theta$, $g(\mathcal{C}(\mathcal{M}, \theta)) = F(\theta)$. ◀

In Defintion 9.1, if the condition $g(\mathcal{C}(\mathcal{M}, \theta)) = F(\theta) \; \forall \theta$ is satisfied for each equilibrium of the game associated with the mechanism, then the implementation is called *strong implementation*. In Example 9.3, Bayesian game associated with the mechanism IM1 has two Bayesian-Nash equilibria, but the desired procurement function is implemented in one of them; hence it is called *weak implementation*.

We know that implementation in dominant strategy equilibrium is the most preferred implementation—dominant strategy is the most desired solution concept $\mathcal{C}$ in Definition 9.1.

In Sect. 9.1.2, we cited Hurwicz (1960, 1972) and mentioned that there may be many possible mechanisms that can achieve the desired goals. For example, for the vaccine procurement problem, we have discussed multiple mechanisms that can implement the desired procurement function. We also posed—*can we design an optimal mechanism that can achieve the desired goals?* One of the most important results in mechanism design, called the *revelation principle*, guides us in answering the question. The revelation principle, originally proposed by Gibbard (1973), tells us to restrict our search to *direct mechanisms with a truth-revealing equilibrium*. In a direct mechanism, the message or strategy set for agent i is the private information set. For the vaccine procurement problem, any mechanism with $M_i = C_i = \{1, 2\}$ is a direct mechanism. Sealed-bid auctions are direct mechanisms as shown in Fig. 9.16, where $\mathbf{x}(\mathbf{b})$ is the allocation rule based on the submitted bids and $\mathbf{p}(\mathbf{b})$ is the payment rule based on the submitted bids.

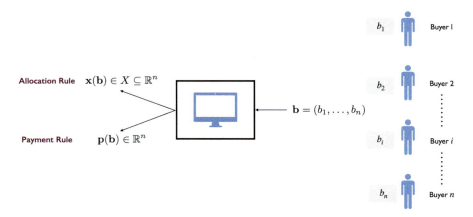

Fig. 9.16 Sealed-bid auctions as direct mechanisms

It is important to note that FPSB auction, being a sealed-bid auction, is a direct mechanism; however, it does not have truth-revealing equilibrium (refer Fig. 9.14). SPSB auction is a direct mechanism with truth-revealing dominant strategy equilibrium (refer Fig. 9.15).

9.2.1 Revelation Principle

We first state the results.

Theorem 9.1 (Revelation Principle for Dominant Strategy Equilibrium) *The outcome that can be achieved in the dominant strategy equilibrium of any mechanism can also be achieved by a direct mechanism with a truth-revealing dominant strategy equilibrium.* ◄

Theorem 9.2 (Revelation Principle for Bayesian-Nash Equilibrium) *The outcome that can be achieved in Bayesian-Nash equilibrium of any mechanism can also be achieved by a direct mechanism with a truth-revealing Bayesian-Nash equilibrium.* ◄

We discuss the proof idea using first-price sealed-bid (FPSB) auction. In Example 9.8, we mentioned that in FPSB auction, the good is auctioned to the bidder with the highest valuation; however, FPSB auction is not truth revealing in Bayesian-Nash equilibrium.

Let us construct a *truth-revealing direct mechanism* with FPSB auction embedded into it, as shown in Fig. 9.17. In truth-revealing direct mechanism, each buyer has a *disinterested proxy buyer*[12] who submits an optimal bid on behalf of the original buyer based on the private valuation of the original buyer. Before the auction, each buyer messages his valuation to the proxy buyer. If the other buyer messages his true valuation, then it is optimal for a buyer to message his true valuation to his proxy buyer because messaging any other valuation leads to a suboptimal bid by his proxy buyer. Hence, truthful revelation forms Bayesian-Nash equilibrium in a truth-revealing direct mechanism with the same outcome (allocation and prices) that can be achieved in Bayesian-Nash equilibrium in FPSB auction. This is the statement of Theorem 9.2.

The revelation principle tells us to forgo complex mechanisms to achieve an outcome and focus on truth-revealing direct mechanisms. However, as shown in Fig. 9.17, the task of computing the optimal strategy shifts from the agents to the mechanism, which enhances the computational burden for the mechanism. We know that in SPSB auction, bidding true valuation is a dominant strategy. If we consider a truth-revealing direct mechanism with SPSB auction as shown in Fig. 9.18, the computational burden does not increase as proxy buyers just bid the valuation messaged by the respective original buyers. Each original buyer messages true valuation in

---

[12] eBay uses proxy bidding in auctions (Source: eBay.com).

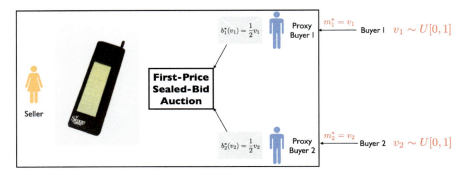

Fig. 9.17 Revelation principle: truth-revealing direct mechanism with FPSB auction

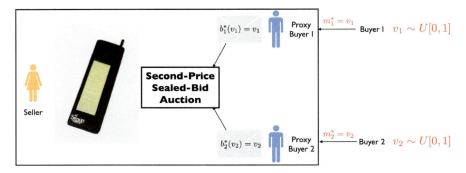

Fig. 9.18 Revelation principle: truth-revealing direct mechanism with SPSB auction

dominant strategies; otherwise, it leads to suboptimal bidding by the proxy bidder. This discussion proves Theorem 9.1. We recommend Shoham and Leyton-Brown (2009) for further reading on the revelation principle.

The proxy buyers are just communication conduits in Fig. 9.18, and *SPSB auction is itself a direct mechanism with truth-revealing dominant strategy equilibrium.* Furthermore, SPSB auction has limited computational requirements. We discuss the goodness properties of SPSB auction in Sect. 9.3.

9.2.2 Procurement Problem Revisited

The mechanisms used for vaccine procurement in Examples 9.2–9.5 are direct mechanisms with different payment schemes.

Example 9.9 (*Optimal Direct Procurement Mechanism*) We use the wisdom of the revelation principle of restricting to direct mechanisms and design an *optimal direct*

9.2 Mechanism Design: Formalism

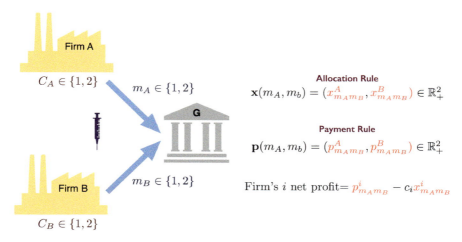

Fig. 9.19 Framework for optimal direct procurement mechanism

mechanism that implements the desired procurement function (refer Example 9.1) in dominant strategies.

The framework for optimal direct procurement mechanism $(\mathbf{x}, \mathbf{p})$ is shown in Fig. 9.19. $\mathbf{x}(m_A, m_B) = (x^A_{m_A m_B}, x^B_{m_A m_B})$ is the allocation (or procurement) rule that allocates $x^A_{m_A m_B}$ and $x^B_{m_A m_B}$ doses of vaccine to Firm A and Firm B for procurement, respectively, based on the messages m_A and m_B. For example, if $m_A = 1$ and $m_B = 1$, then $\mathbf{x}(1, 1) = (x^A_{11}, x^B_{11})$. Similarly, $\mathbf{p}(m_A, m_B) = (p^A_{m_A m_B}, p^B_{m_A m_B})$ is the payment rule that provides total payment $p^A_{m_A m_B}$ and $p^B_{m_A m_B}$ to Firm A and Firm B, respectively, based on the messages m_A and m_B. Firm i's net profit function is $p^i_{m_A m_B} - c_i x^i_{m_A m_B}$ is based on the allocation rule, the payment rule, and its cost. The net profit function satisfies the quasilinear preference property (Sect. 9.1.2).

We formulate the problem of designing the optimal direct procurement mechanism as an optimization problem that computes the optimal payment rule. *The optimal payment rule $\mathbf{p}$ provides the incentives that are compatible with truth revelation in dominant strategies so that the outcome from the allocation rule $\mathbf{x}$ matches with desired procurement function.*[13]

In the optimization formulation, the first set of constraints is the *individual rationality* (IR) constraints (or *participation constraints*) that ensure nonnegative profit for each firm for truthfully messaging its true cost, independent of the message of the other firm. Equation IR-A are the individual rationality constraints for Firm A. The first two constraints ensure nonnegative profit for Firm A by messaging its true cost as 1, independent of whether Firm B messages its cost as 1 (the first constraint)

---

[13] We can use the desired procurement function as the allocation rule while designing an optimal direct mechanism. This is similar to use the desired social function as the outcome function in Definition 9.1 for direct mechanisms. We recommend Mas-Colell et al. (1995, p. 868) to read more on this aspect of direct mechanisms.

or 2 (the second constraint). Similarly, IR-B are the individual rationality constraints for Firm B.

$$p_{11}^A - 1 \cdot x_{11}^A \geq 0$$
$$p_{12}^A - 1 \cdot x_{12}^A \geq 0$$
$$p_{21}^A - 2 \cdot x_{21}^A \geq 0 \qquad \text{(IR-A)}$$
$$p_{22}^A - 2 \cdot x_{22}^A \geq 0$$

$$p_{11}^B - 1 \cdot x_{11}^B \geq 0$$
$$p_{12}^B - 2 \cdot x_{12}^B \geq 0$$
$$p_{21}^B - 1 \cdot x_{21}^B \geq 0 \qquad \text{(IR-B)}$$
$$p_{22}^B - 2 \cdot x_{22}^B \geq 0$$

The second set of constraints is the *incentive compatibility* (IC) *constraints* that ensure each firm has the incentives compatible with messaging its true cost, independent of the message of the other firm. Equation IC-A are the incentive compatibility constraints for Firm A. The first two constraints ensure higher profit for Firm A by messaging its true cost as 1 rather than misreporting the cost as 2, independent of whether Firm B messages its cost as 1 (the first constraint) or 2 (the second constraint). Similarly, equation IC-B are the incentive-compatibility constraints for Firm B.

$$p_{11}^A - 1 \cdot x_{11}^A \geq p_{21}^A - 1 \cdot x_{21}^A$$
$$p_{12}^A - 1 \cdot x_{12}^A \geq p_{22}^A - 1 \cdot x_{22}^A$$
$$p_{21}^A - 2 \cdot x_{21}^A \geq p_{11}^A - 2 \cdot x_{11}^A \qquad \text{(IC-A)}$$
$$p_{22}^A - 2 \cdot x_{22}^A \geq p_{12}^A - 2 \cdot x_{12}^A$$

$$p_{11}^B - 1 \cdot x_{11}^B \geq p_{12}^B - 1 \cdot x_{12}^B$$
$$p_{21}^B - 1 \cdot x_{21}^B \geq p_{22}^B - 1 \cdot x_{22}^B$$
$$p_{12}^B - 2 \cdot x_{12}^B \geq p_{11}^B - 2 \cdot x_{11}^B \qquad \text{(IC-B)}$$
$$p_{22}^B - 2 \cdot x_{22}^B \geq p_{21}^B - 2 \cdot x_{21}^B$$

Equations IR-A, IR-B, IC-A, and IC-B guarantee participation and truthful revelation of its true cost as dominant strategies for each firm.

The next set of constraints is the *quantity constraints* (QC) that ensure the government can procure 100 million doses of the vaccine in each possible state.

9.2 Mechanism Design: Formalism

$$x_{11}^A + x_{11}^B \geq 100$$
$$x_{12}^A + x_{12}^B \geq 100$$
$$x_{21}^A + x_{21}^B \geq 100 \qquad \text{(QC)}$$
$$x_{22}^A + x_{22}^B \geq 100$$

The last set of constraints is the *fairness constraints* (FC) that provide a guarantee to the firms that when they message the same cost, the quantity procured from each firm is equal.

$$x_{11}^A = x_{11}^B$$
$$x_{22}^A = x_{22}^B \qquad \text{(FC)}$$

In this optimization formulation, there are sixteen decision variables, four for each state, and there are four possible states:

$$x_{m_A m_B}^A, x_{m_A m_B}^B, p_{m_A m_B}^A, p_{m_A m_B}^B$$

The decision variables satisfy nonnegativity constraints (NC) for each possible state.

$$x_{m_A m_B}^A \geq 0$$
$$x_{m_A m_B}^B \geq 0$$
$$p_{m_A m_B}^A \geq 0 \qquad \text{(NC)}$$
$$p_{m_A m_B}^B \geq 0$$

The goal of the government is to implement the desired procurement function to minimize the expected procurement cost of 100 million vaccine doses. The objective function of the government in the optimal direct procurement mechanism is

$$Z = \min \frac{1}{4}(p_{11}^A + p_{11}^B) + \frac{1}{4}(p_{12}^A + p_{12}^B) + \frac{1}{4}(p_{21}^A + p_{21}^B) + \frac{1}{4}(p_{22}^A + p_{22}^B)$$
$$\text{(Obj. Func.)}$$

The objective function minimizes the procurement cost to the government in each possible state, and each state can occur with probability $\frac{1}{4}$. The objective function (Obj. Func.) subject to the constraints IR-A, IR-B, IC-A, IC-B, QC, FC, and NC form a linear programming problem that can be easily solved[14] to find the optimal solution. The optimal solution[15] given by the solver is

---

[14] We use Microsoft® Excel for Mac Solver to solve the linear programming problem.

[15] The optimal solution is the same if we replace the allocation rule with the desired procurement function in the formulation, as discussed in Footnote 13 on Sect. 9.2.2, and compute the optimal payment rule.

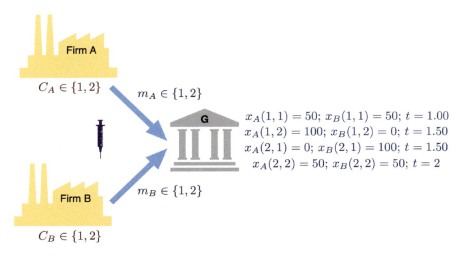

Fig. 9.20 Vaccine procurement—optimal direct mechanism

$$x^A_{11} = 50;\ x^A_{12} = 100;\ x^A_{21} = 0;\ x^A_{22} = 50$$
$$x^B_{11} = 50;\ x^B_{12} = 0;\ x^B_{21} = 100;\ x^B_{22} = 50$$
$$p^A_{11} = 50;\ p^A_{12} = 150;\ p^A_{21} = 0;\ p^A_{22} = 100$$
$$p^B_{11} = 50;\ p^B_{12} = 0;\ p^B_{21} = 150;\ p^B_{22} = 100$$
$$Z = 150\ \text{million dollars}$$

It can be checked that the allocation in the optimal solution matches with the desired procurement function as the optimal payment rule provides the incentives compatible with participation and truth revelation in dominant strategies in the mechanism. Using the revelation principle, no other mechanism can do better in implementing the desired procurement function in dominant strategies.

The optimal direct procurement mechanism is shown in Fig. 9.20. The optimal price per dose is computed using the optimal solution, for example, $t = \frac{150}{100} = 1.50$ dollars per dose for $x^A_{12} = 100$ and $p^A_{12} = 150$. The payment scheme in the optimal mechanism is close to the payment scheme in IM3 in Example 9.5.[16] The difference is due to strictly dominant strategy implementation in IM3 compared to weak dominant strategy implementation in the optimal mechanism. As mentioned at the end of Example 9.5, the optimal mechanism reduces the expected cost until the incentives are right for messaging the true cost as a dominant strategy for each firm. ◁

---

[16] The payment scheme guessed by the consultant in IM3 is close to the optimal payment scheme.

9.3 Auctions

We discussed the sealed-bid auctions in Example 9.8. The auctions are mechanism design problems, and the modern auction theory uses the language of game theory. Auctions have been used widely in different contexts,[17] some of them are shown in Fig. 9.11. In this section, we discuss some important results in auction theory. Before we discuss further, let us look at other famous auction formats.[18]

- **English Auctions:** English auction is one of the most popular auction formats in which the bidders announce successively higher bids, and the winner of the auction is the bidder with the highest prevailing bid when the auction stops. The winner must pay his own bid. The bid increment and the stopping criterion are part of the rules specified by the auctioneer. The stopping criterion can be the fixed time in which no new bids are made.
- **Japanese Auctions:** In Japanese auction, the auctioneer successively increases prices, and the bidders have to signal their willingness to buy at the prevailing price by pressing a button.[19] A bidder drops out when he stops pressing the button and cannot reenter. The auction stops when there is only one bidder remains active. The active bidder is the winner and must pay the prevailing price.
- **Dutch Auctions:** In a Dutch auction, the auctioneer starts from a high price and successively decreases prices. The auction stops when a bidder presses a button to announce his willingness to buy at the prevailing price. This bidder is the winner and must pay the prevailing price. Dutch auctions are often used for auctioning perishable items.

As evident, the above auction formats are *not* direct mechanisms, and we know from the revelation principle that we can confine our attention to truth-revealing direct mechanisms. Second-price sealed-bid (SPSB) auction is one such mechanism that is truth revealing in dominant strategies. In the following sections, we discuss SPSB auctions and other auction mechanisms akin to SPSB auctions for different settings.

9.3.1 Desiderata for Auctions

In the remaining part of the chapter, we restrict our attention to direct auction mechanisms, especially the SPSB auction. As shown in Fig. 9.21, a direct auction mechanism has the following ingredients:

i. **Bid Format:** Bid format is the form of the bids. A bid can have *single attribute* like valuation, or *multiple attributes* like valuation and quantity needed. The

---

[17] Bertsimas and Tsitsiklis (1997) discuss auctions for solving network flow problems. Hall and McMullen (2004) discuss auctions for multisensor data fusion.

[18] Paul Milgrom and Robert Wilson shared the 2020 Nobel Prize in Economic Sciences "for improvements to auction theory and inventions of new auction formats".

[19] Japanese auction is also known as *Japanese button auction*.

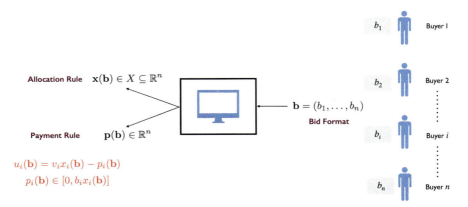

Fig. 9.21 Direct auction mechanism

bid format shown in Fig. 9.21 has a single attribute. Combinatorial auctions use a bid format with multiple attributes, and analyzing them is computationally challenging. Combinatorial auctions are discussed in Sect. 9.3.4.

ii. **Allocation Rule:** The allocation rule gives the outcome based on the submitted bids. For example, in a single-item auction, $\mathbf{x}(\mathbf{b}) = (x_1, \ldots, x_i, \ldots, x_n)$ with only $x_i = 1$ if bidder i wins the auction.

iii. **Payment Rule:** The payment rule gives the payment to be made based on the submitted bids. We consider the payment rules that restrict $p_i(\mathbf{b})$ to the interval $[0, b_i x_i(\mathbf{b})]$, which means that a truthful bidder always gets a nonnegative utility $u_i(\mathbf{b})$. This is similar to the participation constraints in Example 9.9. Furthermore, utility function $u_i(\mathbf{b})$ is *quasilinear* in nature.[20]

iv. **Goals:** When the auctioneer's goal is to auction such the total social welfare is maximized, then the auctions are called *efficient* (or *welfare-maximizing*) *auctions*. In a single-item setting, SPSB auction is an efficient auction as it sells the item to the bidder with the highest valuation. When the auctioneer's goal is to auction such the expected revenue is maximized, then the auctions are called *optimal* (or *revenue-maximizing*) *auctions*. We discuss efficient auctions in Sect. 9.3.2 and optimal auctions in Sect. 9.3.3.

9.3.2 Efficient Auctions

An efficient auction maximizes the total social welfare. If the goal of the auctioneer is to design *direct efficient* auction, then an *efficient allocation rule* for the auctioneer is $\mathbf{x}(\mathbf{v})$ that maximize $\sum_i v_i x_i$, where $\mathbf{v} = (v_1, \ldots, v_n)$ is the true valuation profile

---

[20] Recall the Gibbard–Satterthwaite theorem.

9.3 Auctions 195

that is not known to the auctioneer. For a single-item auction, $\mathbf{x}(\mathbf{v})$ has $x_i = 1$ only for the bidder that has the highest valuation; otherwise, $x_i = 0$.

Considering the advantages of dominant strategy implementation discussed in Sect. 9.1.2, the auctioneer wants to design an auction that implements the efficient allocation rule in dominant strategies—computing the payment rule $\mathbf{p}$ to provide the incentives that are compatible with truth revelation in dominant strategies so that the outcome from the allocation rule of the auction $\mathbf{x}(\mathbf{b})$ matches with the efficient allocation rule $\mathbf{x}(\mathbf{v})$.[21]

For single-item auction settings, we know that SPSB auction implements the efficient allocation rule in dominant strategies (refer Example 9.8). SPSB auction also ensures participation as it provides nonnegative utility to the truthful bidders (refer to the discussion on payment rule in Sect. 9.3.1). Furthermore, as discussed in Sect. 9.2.1, SPSB auction has limited computational requirements.

What is the optimal payment rule $\mathbf{p}$ that implements the efficient allocation rule[22] in dominant strategies for the class of *single-attribute auctions*?[23] This question is answered by *Myerson's Lemma* (Myerson, 1981), which is one of the fundamental results in auction theory.

9.3.2.1 Myerson Lemma for Efficient Auctions

For single-attribute auctions, Myerson's lemma tells us that an efficient allocation rule is implementable in dominant strategies if and only if it is *monotonic* and there is a *unique* optimal payment rule that implements the monotonic efficient allocation rule.

Definition 9.2 (*Monotonic Efficient Allocation Rule*) An efficient allocation rule is monotonic if, for each bidder i and bids $\mathbf{b}_{-i}$ of the other bidders, $x_i(m_i, \mathbf{b}_{-i})$ is nondecreasing with an increase in m_i. ◀

It means that, in a single-attribute auction setting with a monotonic allocation rule, a bidder can improve his allocation by increasing his bids. In the vaccine procurement problem (a reverse auction setting) in Example 9.9, it can be verified the desired procurement function as a function of messaged cost is monotonic.

Theorem 9.3 (Myerson's Lemma) *An efficient allocation rule is implementable in dominant strategies if and only if it is monotonic, and there is a unique optimal payment rule that implements the monotonic efficient allocation rule in dominant strategies, which is given by*

---

[21] The mechanisms that provide incentives compatible with truth revelation in dominant strategies are called *dominant strategy incentive compatible* (**DSIC**).

[22] There are single-attribute auctions like *knapsack auctions* where it is computationally hard to compute the efficient allocation (Roughgarden, 2016). Such auctions are part of *algorithmic mechanism design*. We recommend Nisan (2015) for a survey on algorithmic mechanism design.

[23] We posed a similar question in Example 9.9 for the vaccine procurement problem.

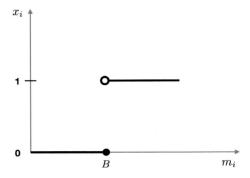

Fig. 9.22 x_i as a function of m_i (Adapted from Roughgarden, 2016)

$$p_i(b_i, \mathbf{b}_{-i}) = \sum_k m_i^k \cdot \triangle x_i^k$$
$$p_i(0, \mathbf{b}_{-i}) = 0$$
$$m_i^k \in [0, b_i]$$

◂

In Theorem 9.3, $m_i^k \in [0, b_i]$ is a *pivotal bid* that changes the allocation x_i of bidder i. For a monotonic allocation rule, x_i always increases at pivotal bids, and k is the number of pivotal points. The unique optimal payment rule is the aggregate of the product of the pivotal bid m_i^k and the increment in the allocation $\triangle x_i^k$ at m_i^k, with the boundary condition $p_i(0, \mathbf{b}_{-i}) = 0$.

For a single-item auction, the allocation x_i is plotted as a function of m_i as shown in Fig. 9.22.

$B = \max_{j \neq i} b_j$ is the highest bid among all the other bidders, excluding bidder i. In a single-item auction, there is only one pivotal bid ($k = 1$) for bidder i, $m_i^1 = B$ when x_i changes from 0 to 1 ($\triangle x_i^1 = 1$). Using the unique payment rule $p_i(b_i, \mathbf{b}_{-i}) = m_i^1 \cdot \triangle x_i^1 = B$. If bidder i is the winner of a single-item auction, then using the unique payment rule, he pays the second-highest bid. The optimal payment rule matches with the payment rule in SPSB auction. In other words, the SPSB auction uses the unique optimal payment rule.

Myerson's lemma is a powerful result for single-attribute auction design[24] *One important inference is that any other payment rule cannot implement a monotonic efficient allocation rule in dominant strategies.*[25]

---

[24] For differentiable x_i, the unique payment rule is

$$p_i(b_i, \mathbf{b}_{-i}) = \int_0^{b_i} m_i \left(\frac{d}{dm_i} x_i(m_i, \mathbf{b}_{-i}) \right) dm_i$$
$$p_i(0, \mathbf{b}_{-i}) = 0$$

[25] Generalizes second-price (GSP) auction is not truthful (refer Sect. 10.1).

9.3 Auctions

Proof of Theorem 9.3 Consider an efficient allocation rule $\mathbf{x}(\mathbf{b})$, which the auctioneer desires to implement in dominant strategies. For notational brevity, we use $x_i(m_i)$ and $p_i(m_i)$ for $x_i(m_i, \mathbf{b}_{-i})$ and $p_i(m_i, \mathbf{b}_{-i})$, respectively.

For each bid profile of the other agents $\mathbf{b}_{-i}$, and if bidder i has two possible valuations m_i and m_i', then the incentive-compatibility constraints[26] (IC-Myerson's Lemma) that ensure truth revelation in dominant strategies are

$$m_i \cdot x_i(m_i) - p_i(m_i) \geq m_i \cdot x_i(m_i') - p_i(m_i')$$
$$m_i' \cdot x_i(m_i') - p_i(m_i') \geq m_i' \cdot x_i(m_i) - p_i(m_i) \qquad \text{(IC-Myerson's Lemma)}$$

Let $m_i' = m_i + \triangle m_i$ with $\triangle m_i > 0$, and substituting m_i' in the above constraints and rearranging results in

$$(m_i + \triangle m_i)(x_i(m_i + \triangle m_i) - x_i(m_i))$$
$$\geq p_i(m_i + \triangle m_i) - p_i(m_i) \geq m_i(x_i(m_i + \triangle m_i) - x_i(m_i)) \qquad (\spadesuit)$$

This means

$$(m_i + \triangle m_i)(x_i(m_i + \triangle m_i) - x_i(m_i)) \geq m_i(x_i(m_i + \triangle m_i) - x_i(m_i))$$

Further rearrangement results in

$$\triangle m_i(x_i(m_i + \triangle m_i) - x_i(m_i)) \geq 0$$

For $\triangle m_i > 0$, the above condition can only be satisfied if x_i is nondecreasing with an increase in m_i, or the efficient allocation rule $\mathbf{x}(\mathbf{b})$ is monotonic.

Nonmonotonic allocation rules violate the incentive-compatibility constraints for dominant strategy implementation and hence cannot be implemented in dominant strategies. This completes the proof of the first part of the theorem.

In Eq. ($\spadesuit$), when $\triangle m_i \longrightarrow 0$, both the left-hand and the right-hand tends to $m_i \cdot \triangle x_i$, and this results in

$$\triangle p_i = m_i \cdot \triangle x_i$$

Using $p_i(0) = 0$ and $m_i \in [0, b_i]$ along with the above equation leads to

$$p_i(b_i, \mathbf{b}_{-i}) = \sum_k m_i^k \cdot \triangle x_i^k$$

where $m_i^k \in [0, b_i]$ is a *pivotal bid* that increments the allocation x_i of bidder i. This is the only payment rule that is consistent with incentive compatibility in dominant strategies for implementing monotonic efficient allocation rules in single-attribute auctions. This proves the second part of the theorem. $\qquad \square$

---

[26] Recall the IC constraints in Example 9.9.

9.3.2.2 Revenue Equivalence

We have discussed various types of auction formats, such as sealed-bid auctions, English auctions, Dutch auctions, and Japanese auctions, at the beginning of this section. In that context, the question arises *"Which format should the auctioneer follow to maximize its revenue?"* In this section, we will look closely at this part and provide a strong result of auction theory.

In Sect. 9.2.1, we have seen that the first-price auction and the second price have different payment rules. Therefore a natural question arises for an auctioneer whether to choose first-price or second-price auction. In reality, the expected value that the auctioneer would get in both these auctions is the same, which is known as the *revenue equivalence theorem*. We will prove this using a simple example for auctions where the private valuations are uniformly distributed. In any auction, the bidders have a private valuation and bid based on that. For a bidder i, its *strategy* (s_i) is to transform the private valuation to a bid (b_i) through some mapping. Therefore a strategy can be considered as a function that maps valuation to bid. We assume that the strategy is symmetric across all bidders, i.e., they use the same function to transform their valuation into a bid. This bid function is always monotonic in nature.

Let us consider an auction with n bidders, and their valuation is uniformly distributed between [0, 1]. The valuation profile $\mathbf{v} = (v_1, \ldots, v_n)$ is private knowledge, and the auctioneer wants to know the expected value of the highest valuation. When we treat the valuation profile as the series of independent draws from a distribution $F(.)$, we can consider the ranking of these values such as $v_n^{[1]} = \max\{v_1, v_2, \ldots, v_n\}$. This is the first-order statistic. Similarly, the second-order statistic gives the next highest value, and so on. We understand that the cumulative density function $F(v)$ gives the probability of having all the valuations less than v, i.e.,

$$F(v) = Prob(v_i \leq v \ \text{ and } \ v_2 \leq v \ldots \ \text{ and } \ v_n \leq v) \tag{9.1}$$

For a single bidder $F^i(v)$ can easily be calculated as $\frac{v-0}{1-0} = v$. As each bidder's valuation is independent, the CDF for all the bidders will be v^n. The probability density function can easily be calculated from the CDF by taking a derivative with respect to v, and that would give the PDF as nv^{n-1}. Now we can calculate the expectation of the highest valuation $(v_n^{[1]})$ in the valuation profile from this by the following method;

$$E(v_n^{[1]}) = \int_0^1 v \times nv^{n-1} dv = n \int_0^1 v^n dv \tag{9.2}$$

$$= n \left[\frac{v^{n+1}}{n+1} \right] = \frac{n}{n+1} \tag{9.3}$$

Similarly, it can be calculated that the expectation of the second-highest valuation is $E(v_2) = \frac{n-1}{n+1}$. In the second-price auction, as the payment made by the winner is the second-highest value, $\frac{n-1}{n+1}$ is the expected revenue for the auctioneer. Let us

9.3 Auctions

calculate the expected value for a first-price auction in this scenario. For a first-price auction, when there are n bidders, and valuations are uniformly distributed, the bidder's optimal bidding strategy with a valuation v is given as,

$$s(v) = \frac{n-1}{n}v \tag{9.4}$$

Considering each bidder as a rational agent and bidding optimally, we can say that the payment that the auctioneer would get is the above expression for the item. Equation 9.2 gives us the expected value of the highest valuation, v. As, in the first-price auction, the highest bidder gives its own valuation, the expected revenue that the seller can get can be calculated as:

$$\frac{n-1}{n}v = \frac{n-1}{n} \times \frac{n}{n+1} = \frac{n-1}{n+1} \tag{9.5}$$

This expression is the same as the expected revenue for the seller in the second-price auction. This gives us the result of the revenue equivalence theorem.

Revenue Equivalence Theorem: For any auction format having n bidders having the private valuation are identically and independently distributed, and the bidder with the highest valuation is declared as the winner yields **same expected revenue** for the auctioneer irrespective of the format it chooses. In other words, it can be stated that under certain general conditions, all auction formats result in the same expected revenue. Readers can check for other formats of actions by following the same procedure as shown above.

9.3.3 Optimal Auctions

We discussed efficient auctions in the previous section, where the auctioneer's goal is to maximize the total social welfare. Myerson's lemma tells us that the optimal payment rule, which implements the efficient allocation rule in dominant strategies, is unique. For efficient auctions, the role of the optimal payment rule is to provide incentives to implement the efficient allocation rule in dominant strategies rather than the revenue considerations.

Implementing efficient allocation rules is relevant in public sector settings like in spectrum auctions. However, in numerous business settings, the auctioneer's goal is to maximize the expected revenue, and the auctions in these settings are called *optimal* (or *revenue-maximizing*) *auctions*. Designing optimal auctions is more complex than efficient auctions as optimal auctions are not *detail-free mechanisms*. In Sect. 9.3.2, efficient auctions are detail-free as the information about the distribution of the bidders' valuation is not needed while designing them. Designing optimal auctions needs such information.

9.3.3.1 Optimal Selling Mechanism with Single Buyer

Consider a single-item single-buyer setting[27] in which the buyer has a private valuation v with F as the CDF of v. If the CDF is *known* to the seller, the seller can design a "take-it-or-leave-it" mechanism with a posted price r to maximize the expected revenue.

For a posted price r, the item is sold when $r \leq v$, and this happens with probability $(1 - F(r))$; otherwise, the item remains unsold. $(1 - F(r))$ is similar to a typical demand function in elementary microeconomics—as r decreases, the likelihood of the sale increases. The expected revenue to the seller is

$$R = r \cdot (1 - F(r))$$

The expected revenue R is maximized when the marginal revenue $MR = 0$, and the optimal posted price r^* satisfies the condition

$$MR \equiv r^* - \frac{(1 - F(r^*))}{f(r^*)} = 0$$

f is the PDF of F. The optimal posted price r^* determines the *optimal allocation rule*[28]—if $v \geq r^*$, the item is bought by the buyer; otherwise, it remains with the seller.

The optimal selling mechanism is *truth revealing*. Given the optimal allocation rule as a function of the optimal posted price r^*, if the buyer is asked to reveal his valuation, the buyer reveals its truthfully as the allocation is independent of v. However, the optimal selling mechanism is not *detail-free* as the optimal selling price depends on the distribution of valuation of the buyer.

Furthermore, we know from elementary microeconomics that a revenue-maximizing seller operates in the demand range where the marginal revenue is nonnegative. In the optimal selling mechanism, if the buyer buys at the posted price r^*, then $v \geq r^*$ and $MR(v) \geq 0$, as $MR(r^*) = 0$. In other words, in the single-item single-buyer setting, the optimal allocation rule assigns the item to the buyer if the buyer has nonnegative marginal revenue.

Also, the condition $v \geq r^*$ and $MR(v) \geq 0$ implies that marginal revenue function MR is a *nondecreasing function* in valuation. This is possible if the valuation distribution satisfies the *monotone hazard rate condition* (MHR Condition). Many common distributions, including uniform distribution, satisfy the MHR condition.

$$\text{Hazard rate } \frac{f(v)}{1 - F(v)} \text{ is nondecreasing in } v \qquad \text{(MHR Condition)}$$

---

[27] In microeconomics, the single-buyer-single-seller setting is called *bilateral monopoly*.

[28] Optimal allocation rule maximizes the expected revenue. Recall the efficient allocation rule that maximizes the total social welfare.

9.3 Auctions

9.3.3.2 Optimal Auctions with Multiple Buyers

As mentioned earlier, designing optimal auctions with multiple bidders is more complex than efficient auctions. In optimal auctions, the auctioneer's goal is to maximize the expected revenue, with valuations being private information of the bidders. We discussed the significance of dominant strategies in mechanism design. Hence, the auctioneer still searches for mechanisms that maximize the expected revenue, with bidders truthfully revealing their private valuations in dominant strategies. Using the revelation principle, the auctioneer can restrict the search to *direct mechanisms*. In other words, designing an optimal auction means that *the auctioneer wants to design a direct auction mechanism that implements the optimal allocation rule*[29] *in dominant strategies*.

We know that, for efficient auctions, Myerson's lemma is an important result—only the monotonic efficient allocation rule can be implemented in dominant strategies using the unique payment rule. Can a revenue-maximizing auctioneer show the equivalence of the optimal allocation rule to an efficient allocation rule and then use Myerson's lemma to design optimal auctions? The answer is "yes", and we need some wisdom from the optimal selling mechanism (Sect. 9.3.3.1) to show the equivalence.

In efficient auctions, the auctioneer's goal is to implement the efficient allocation rule $\mathbf{x}(\mathbf{v})$ that maximizes $\sum_i v_i x_i$, where $\mathbf{v} = (v_1, \ldots, v_n)$ is the true valuation profile that is not known to the auctioneer.

In single-item single-buyer setting, to maximize the expected revenue, the seller sets the posted price r^* at which marginal revenue is zero ($MR = 0$), and the item is sold if $v \geq r^*$, and $MR(v) \geq 0$.[30] The optimal allocation rule assigns the item to the buyer if the buyer has nonnegative marginal revenue. This can be extended to the multiple-bidders setting where the revenue-maximizing auctioneer's goal is to implement the optimal allocation rule $\mathbf{y}(\mathbf{v})$ that maximizes $\sum_i MR_i(v_i) \cdot y_i$ among the bidders with $MR_i(v_i) \geq 0$.[31] If there is no bidder with $MR_i(v_i) \geq 0$, then the items remain with the auctioneer.

For single-item auctions, the optimal allocation rule maximizes the expected revenue by assigning the item to the bidder with the highest marginal revenue, subject to the condition that the marginal revenue of the winning bidder is nonnegative. This is similar to single-item efficient auctions, where the efficient allocation rule assigns the item to the bidder with the highest valuation. This equivalence between the optimal allocation rule and the efficient allocation rule allows us to use Myerson's lemma to design optimal auctions; hence, marginal revenue is called *virtual valuation* γ while designing optimal auctions.

$$\gamma_i(v_i) \triangleq MR_i(v_i) = v_i - \frac{(1 - F_i(v_i))}{f_i(v_i)} \tag{$\diamond$}$$

---

[29] Recall Footnote 28 on Sect. 9.3.3.1, the optimal allocation rule maximizes the expected revenue.

[30] Assuming the valuation distribution satisfies the MHR condition.

[31] Recall, a revenue-maximizing auctioneer operates in the demand range where the marginal revenue is nonnegative.

In Eq. $\diamondsuit$, $\gamma_i(v_i)$ is the virtual valuation of the bidder i with private valuation v_i and the valuation CDF F_i. Using virtual valuations, now the goal of the revenue-maximizing auctioneer is to implement the optimal allocation rule $\mathbf{y}(\mathbf{v})$ that maximizes the *total virtual welfare* $\sum_i \gamma_i(v_i) \cdot y_i$ among the bidders with $\gamma_i(v_i) \geq 0$.[32] If there is no bidder with $\gamma_i(v_i) \geq 0$, then the items remain with the auctioneer. The optimal auction setting is now analogous to the efficient auction setting, and if we prove the monotonicity of the optimal allocation rule, Myerson's lemma can guide us in designing truth-revealing optimal auctions.

The optimal allocation rule is *monotonic* if $\gamma_i(v_i)$ is nondecreasing in v_i, which is true if F_i satisfies the (MHR Condition) for each bidder. Using Myerson's lemma, the monotonic optimal allocation rule is implementable in dominant strategies using the unique payment rule (Eq. $\spadesuit$). The implementable optimal allocation rule maximizes the total virtual welfare (or the expected revenue).

$$p_i(b_i, \mathbf{b}_{-i}) = \sum_l m_i^l \cdot \triangle y_i^l \qquad (\spadesuit)$$

where $m_i^l \in [r_i, b_i]$ is a *pivotal bid* that increments the allocation y_i of bidder i. For a monotonic optimal allocation rule, y_i always increases at pivotal bids, and l is the number of pivotal points. r_i is the bidder-specific reserve price at which the virtual valuation of bidder i is zero ($\gamma_i(r_i) = 0$),[33] and the reserve price ensures that nonnegativity of the virtual valuation.

As the payment rule in Eq. $\spadesuit$ ensures truth revelation in dominant strategies, we assume $\mathbf{b} = \mathbf{v}$, and the payment rule can be written as

$$p_i(v_i, \mathbf{v}_{-i}) = \sum_l m_i^l \cdot \triangle y_i^l$$

For a single-item optimal auction, the allocation y_i is plotted as a function of m_i as shown in Fig. 9.23. Bidder i with valuation v_i is the winner, and v_i' is the pivotal bid that satisfies the condition $\gamma_i(v_i') = \gamma_j(v_j)$ with $\gamma_j(v_j) \geq 0$ and $j \neq i$. It means if there is a bidder j with valuation v_j and $\gamma_j(v_j) \geq 0$; and v_i' is the pivotal bid such that $\gamma_i(v_i') = \gamma_j(v_j)$. At the pivotal bid, y_i increments from 0 to 1 ($\triangle y_i = 1$). Using the payment rule, the winner pays $p_i(v_i, \mathbf{v}_{-i}) = v_i'$.

If there is no other bidder j ($j \neq i$) with $\gamma_j(v_j) \geq 0$ and bidder i is the only bidder with $\gamma_i(v_i) \geq 0$, then the winner pays his reserve price $p_i(v_i, \mathbf{v}_{-i}) = r_i$.

The optimal sealed-bid single-item auction protocol with three bidders is shown in Fig. 9.24. The optimal auction is truth revealing in dominant strategies due to the payment rule that satisfies Myerson's lemma; however, it is not detail-free as it needs

---

[32] Virtual valuations can take negative values, unlike the valuations; and, as by definition virtual valuation is equal to marginal revenue, a revenue-maximizing auctioneer does not operate in the range with negative virtual valuation.

[33] Recall the optimal posted price in Sect. 9.3.3.1.

9.3 Auctions

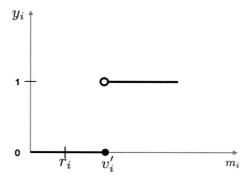

Fig. 9.23 y_i as a function of m_i

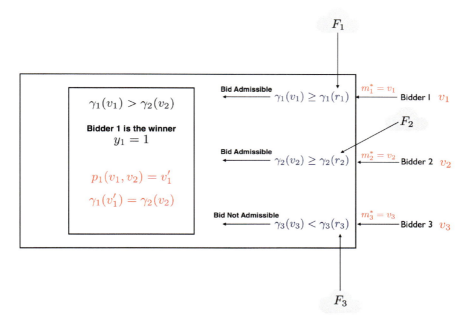

Fig. 9.24 Optimal sealed-bid single-item auction

information about the valuation distribution of each bidder, for example, F_1 of Bidder 1 in Fig. 9.24.

Like efficient auctions in Sect. 9.3.2, optimal auction design discussed in this section is also in *single-attribute* auction settings as Myerson's lemma is applicable only for single-attribute auction settings.

Example 9.10 (*Optimal Symmetric Single-item Auction*) The preceding discussion assumed that the valuation distribution can be different for each bidder. Such auctions are called *asymmetric auctions*. When the valuation distribution F is the same for each bidder, then the virtual valuation function γ is also the same for each bidder. Auctions in such settings are called *symmetric auctions*. Symmetric auctions are operationally simpler than asymmetric auctions as they only need information about the common valuation distribution.

Let us assume F satisfies the MHR condition, then $\gamma(v_i)$ is a nondecreasing function in v_i, and the optimal allocation rule is monotone. Furthermore, the reserve price r^* is the same for each bidder and can be computed using the following condition

$$\gamma(r^*) = r^* - \frac{(1 - F(r^*))}{f(r^*)} = 0$$

In a single-item symmetric auction, the optimal allocation rule is implementable in dominant strategies using the payment rule $p_i(v_i, \mathbf{v}_{-i}) = m_i^1 \cdot \triangle y_i^1$ ($l = 1$ for single-item auction in Eq. ♠). The optimal allocation rule assigns the item to the bidder with the highest virtual valuation. As the virtual valuation function is the same for each bidder, the optimal allocation rule assigns the item to the bidder with the highest valuation, with the winner paying the second-highest valuation. For example, if Bidder 1 and Bidder 2 are the bidders with the highest and second-highest virtual valuations, respectively. Then, Bidder 1 is the winning bidder and pays v_2.[34] *It means the optimal symmetric single-item auction is the second-price auction with an optimal reserve price r^*.* ◁

In the optimal symmetric single-item auction, which is the second-price auction with an optimal reserve price, the reserve price makes the auctioneer act as a bidder with a valuation r^* to increase competition in the auction. Can the auctioneer increase her revenue by using a detail-free second-price auction with an additional bidder to increase competition instead of using second-price auction with an optimal reserve price? The answer is "yes", and the result is due to Bulow and Klemperer (1996). Kirkegaard (2006) provides an intuitive proof of the result by Bulow and Klemperer (1996).

We recommend Myerson (1981), Bulow and Klemperer (1996), Shoham and Leyton-Brown (2009), and Roughgarden (2016) for further reading on optimal auctions.

---

[34] $p_1 = v_1'$ where $\gamma(v_1') = \gamma(v_2)$. Since γ is a nondecreasing function in valuations, it implies $v_1' = v_2$ and $p_1 = v_2$.

9.3.4 Combinatorial Auctions

In the previous sections, we discussed efficient and optimal auctions in single-attribute settings. In this section, we extend the discussion on auctions to multiple-attribute settings where the bid format can have multiple attributes. We confine our discussion to *combinatorial auctions* as a class of multiple-attribute auctions.

In a combinatorial auction, multiple distinct items are auctioned, and bidders can have valuations for different *combinations* of items called *bundles*. If K is the set of items, then the total number of possible bundles is 2^k, where k is the cardinality of set K. Each bidder i has a private *valuation function* $v_i : 2^k \mapsto \mathbb{R}$ that assigns a valuation $v_i(S)$ to bundle $S \subseteq K$. We focus on efficient (or welfare-maximizing) combinatorial auctions[35] where the goal of the auctioneer is to implement an efficient allocation rule $\mathbf{x}(\mathbf{v})$ that maximizes the total social welfare $\sum_{i=1}^{n} v_i(S_i)$ where $S_i \subseteq K$ is the bundle assigned to bidder i. $\mathbf{v} = (v_1, \ldots, v_n)$ is the *valuation function* profile.

To implement the efficient allocation rule, using the revelation principle, the auctioneer can focus on *direct auction mechanisms*. A direct combinatorial auction mechanism is shown in Fig. 9.25. The bid format has multiple attributes—b_i is 2^k-dimensional vector with valuations for each possible bundle. Allocation rule $\mathbf{x}(\mathbf{b})$ gives the outcome $(S_1, \ldots, S_i, \ldots, S_n)$ based on the submitted bids, where $S_i \subseteq K$ is the bundle assigned to bidder i, and $\bigcap_i S_i = \emptyset$. We consider the payment rules that restrict $p_i(\mathbf{b})$ to the interval $[0, b_i(S_i)]$, which means that a truthful bidder always gets a nonnegative utility $u_i(\mathbf{b})$.

There are practical issues in direct combinatorial auctions. As the bid format has multiple attributes—each bid b_i has 2^k dimensions, bidding complexity grows

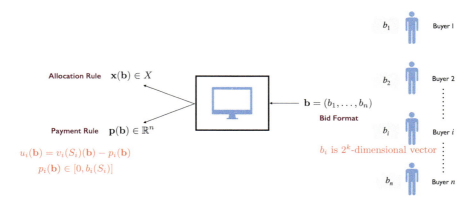

Fig. 9.25 Direct combinatorial auction

---

[35] Revenue-maximizing combinatorial auction design still remains elusive (Dütting et al., 2024).

exponentially with the number of items in K.[36] Furthermorotal social welfare $\sum_{i=1}^{n} v_i(S_i)$ is computationally hard. The problem of computing the efficient allocation is called the *winner determination problem*, which is a class of *set packing problems*. The set packing problems belong to **NP-Complete**[37] class of computational complexity (Shoham & Leyton-Brown, 2009). We are not further discussing these issues with combinatorial auctions as they are part of algorithmic mechanism design, and we focus on implementing the efficient allocation rule.

Considering the advantages of dominant strategy implementation discussed in Sect. 9.1.2, the auctioneer wants to design an auction that implements the efficient allocation rule in dominant strategies—computing the payment rule **p** to provide the incentives that are compatible with truth revelation in dominant strategies so that the outcome from the allocation rule of the auction $\mathbf{x}(\mathbf{b})$ matches with the efficient allocation rule $\mathbf{x}(\mathbf{v})$. This goal is similar to the goal in efficient single-attribute auctions (Sect. 9.3.2), and Myerson's lemma provided the optimal payment rule (Theorem 9.3) that implements the efficient allocation rule in dominant strategies. Unfortunately, Myerson's lemma does not extend to multiple-attribute settings.

However, Myerson's lemma still provides insight to compute the payment rule that implements the efficient allocation in dominant strategies in multiple-attribute settings. In Myerson's lemma, the optimal payment p_i of bidder i is the total valuation loss of the other bidders caused by the presence of bidder i. In a single-item efficient auction setting, the optimal payment of the winning bidder is the valuation of the second-highest bidder. If the winning bidder was not part of the auction, then the second-highest bidder would have won the auction, and hence the winning bidder pays the valuation loss of the second-highest bidder. Other bidders are not affected by the winning bidder's presence in the auction.

For Facebook advertising, this idea is succinctly captured in the following quote by John Hegeman[38]

> If you're an advertiser and you're getting a chance to show your ad, you're going to take away the opportunity from someone else. The price can be determined based on how much value is being displaced from those other people. An advertiser will only win this placement if their ad really is the most relevant, if it really is the best ad to show to this person at this point in time.[39]

Mathematically, the payment rule for bidder i, which captures the valuation loss to other bidders can be written as

$$p_i(\mathbf{b}) = \sum_{j \neq i} b_j(S'_j) - \sum_{j \neq i} b_j(S_j) \qquad (\bigstar)$$

---

[36] The use of *indirect mechanisms* like *simultaneous ascending auctions* (SAA) is prevalent in combinatorial auction settings like *spectrum auctions* to overcome bidding complexity. Haeringer (2018) provides a readable introduction to spectrum auction design.

[37] **NP-Complete** problem can be solved by *magical guessing* as was done by the government's consultant in Example 9.5.

[38] Then Facebook's Chief Economist. Currently, VP, Monetization, Meta.

[39] Source: **Facebook Doesn't Make as Much Money as It Could-On Purpose**, *Wired*, September 21, 2015.

9.3 Auctions

where S_j is the bundle assigned to bidder j using the efficient allocation rule $\mathbf{x}(\mathbf{b})$ that maximized $\sum_{i=1}^{n} b_i(S_i)$. S'_j is the bundle assigned to bidder j using the efficient allocation rule $\mathbf{x}'(\mathbf{b})$ that maximized $\sum_{j \neq i} b_j(S'_j)$. In other words, S'_j is the bundle assigned to bidder j using the efficient allocation rule $\mathbf{x}'(\mathbf{b})$ when bidder i is not present in the auction mechanism. If $S_i \neq \emptyset$ in the efficient allocation rule $\mathbf{x}(\mathbf{b})$, then bidder i's presence leads to valuation loss to other bidders, and $p_i(\mathbf{b})$ is always nonnegative.

The payment rule $\mathbf{p}(\mathbf{b})$ in Eq. ★ is called *Vickrey–Clarke–Groves* (or VCG) *payment rule*, and the efficient allocation rule along with VCG payment rule is called a *Vickrey–Clarke–Groves* (or VCG) *mechanism*.

Theorem 9.4 *In a VCG mechanism, the payment rule in Eq. ★ implements the efficient allocation rule in dominant strategies.*[40] ◄

Proof Given the bids of the other bidders $\mathbf{b}_{-i}$, bidder i's utility (refer Fig. 9.25) is

$$u_i(\mathbf{b}) = v_i(S_i)(\mathbf{b}) - p_i(\mathbf{b})$$

Substituting the VCG payment rule, bidder i's utility is

$$u_i(\mathbf{b}) = \left[v_i(S_i)(\mathbf{b}) + \sum_{j \neq i} b_j(S_j) \right] - \left(\sum_{j \neq i} b_j(S'_j) \right)$$

The term in round brackets does not depend on b_i, and bidder i can ignore its influence when deciding his optimal bid. The term in square brackets depends on b_i. Assume bidder i bids $b_i \neq v_i$, then the auctioneer chooses the allocation $\mathbf{x}(\mathbf{b})$ that maximizes $b_i(S_i) + \sum_{j \neq i} b_j(S_j)$. If bidder i bids truthfully $b_i = v_i$, then the auctioneer chooses the allocation $\mathbf{x}(v_i, \mathbf{b}_{-i})$ that maximizes $v_i(S_i) + \sum_{j \neq i} b_j(S_j)$, which is same as the term in square brackets in bidder i's utility. Hence, by bidding truthfully, bidder i directs the mechanism to choose the allocation that maximizes his own utility. This is true for any $\mathbf{b}_{-i}$, and hence truthful bidding is a dominant strategy for bidder i. The same argument applies for all the bidders, and hence the VCG payment rule implements efficient allocation rule in dominant strategies. □

Example 9.11 Consider a case of having two bidders M and N and two items A and B sold by the seller. The private valuations for both the bidders for the combinations of items are given in the following table.

	M	N
A	12	6
B	7	5
A, B	13	9

From the table we can deduce that $v_M(\{A\}) = 12$, $v_N(\{B\}) = 7$, and $v_M(\{A, B\}) = 13$. For bidder N, this can be written similarly. Four possible assignments are possible, as shown below.

---

[40] A VCG mechanism is dominant strategy incentive compatible (DSIC).

Allocation	M	N	M's valuation	N's valuation	Social value
S_1	A, B		13	0	13
S_2		A, B	0	9	9
S_3	A	B	12	5	17
S_4	B	A	7	6	13

The auction is about implementing the optimal allocation rule, where the objective is to maximize social welfare. From the above table, it is clear that the social value is maximized in the allocation mechanism S_3, which gives the value of 17. Formally, we can write this as

$$\sum_i v_i(S^*(i)) = v_M(S^*(M)) + v_N(S^*(N))$$

$$= v_M(S_3(M)) + v_N(S_3(N))$$
$$= v_M(A) + v_N(B)$$
$$= 12 + 5 = 17$$

The auctioneer computes the optimal allocation based on the bids submitted by the bidders. The next step is to calculate the prices that the bidders have to pay. As mentioned previously, the bidder i pays the difference between the social value of the other bidders at the optimal allocation and the maximum social value of the other bidders when bidder i is not present in the auction. When we consider the above auction setting, if bidder M is not present, then only N is there, and the allocation that gives the highest social value is to allocate both items to N. This gives a value of 9 to N. The presence of M gives the social value of 17, but the value that N gets is 5. Bidder M has to pay the difference between these two values, i.e., $9 - 5 = 4$, to obtain item A. We can also compute the price that N has to pay in a similar fashion, which is $13 - 12 = 1$ for obtaining the item B. ◁

Problems

Problem 9.1 Compute the optimal bidding strategy of a bidder in first-price sealed-bid (FPSB) auction with n bidders and the valuations of all the bidders are distributed uniformly on $[0, 1]$. ◀

Problem 9.2 In a single-item auction, assume the valuations of all the bidders are distributed uniformly on $[0, 1]$. Design *optimal* second-price sealed-bid (SPSB) auction. ◀

Problem 9.3 The government needs to procure 100 million doses of a vaccine. Each firm has linear cost technology with costs that are private information. Each firm's costs are 1, 2, or 3 per unit produced, as shown in the figure. Each cost is equally likely at each firm and independent of the other. Consider the cost probabilities as common knowledge. Suppose the government solicits bids from the firms. If they bid the same amount, the order is split. If one firm bids less, it is given the entire order, and a firm is paid not what it bid but what the other firm bid. Formulate the situation as a Bayesian game and compute its Bayesian-Nash equilibria.

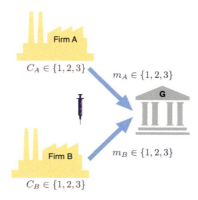

Problem 9.4 The government needs to procure 100 million doses of a vaccine. Each firm has linear cost technology with costs that are private information. Each firm's costs are 1, 2, or 3 per unit produced, as shown in the figure. Each cost is equally likely at each firm and independent of the other. Consider the cost probabilities as common knowledge. Suppose the government solicits bids from the firms. Formulate optimal direct procurement mechanism as a linear program to implement the desired procurement function (refer Example 9.1) in dominant strategies.

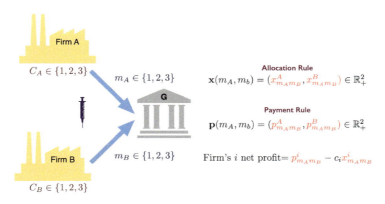

References

Bertsimas, D., & Tsitsiklis, J. N. (1997). *Introduction to linear optimization* (SE - xv, 587 p: illustrations; 24 cm). Belmont, Mass.: Athena Scientific Belmont.

Bulow, J., & Klemperer, P. (1996). Auctions versus negotiations. *American Economic Review, 86*(1), 180–194.

Cassady, R. (2021). *Auctions and auctioneering*. University of California Press.

Davis, R., & Smith, R. G. (1983). Negotiation as a metaphor for distributed problem solving. *Artificial Intelligence, 20*(1), 63–109.

Dütting, P., Feng, Z., Narasimhan, H., Parkes, D. C., & Ravindranath, S. S. (2024). Optimal auctions through deep learning: Advances in differentiable economics. *Journal of the ACM, 71*(1).

Gibbard, A. (1973). Manipulation of voting schemes: A general result. *Econometrica, 41*(4), 587.

Haeringer, G. (2018). *Market design : Auctions and matching*. MIT Press.

Hall, D. L., & McMullen, S. A. H. (2004). *Mathematical techniques in multisensor data fusion* (2nd edn.). Artech House.

Hayek, F. A. (1937). Economics and knowledge. *Economica, 4*, 33–54.

Hayek, F. A. (1945). The use of knowledge in society. *The American Economic Review, 35*(4), 519–530.

Hurwicz, L. (1960). Optimality and information efficiency in resource allocation processes. In K. Arrow, S. Karlin, & P. Suppes (Eds.), *Mathematical methods in the social sciences*. Stanford: Stanford University Press.

Hurwicz, L. (1972). On informationally decentralized systems. In R. Radnar & C. McGuire (Eds.), *Decision and organization*. Amsterdam: North-Holland.

Kelting, M. W. (2009). Tournaments of honor: Jain auctions, gender, and reputation. *History of Religions, 48*(4), 284–308.

Kirkegaard, R. (2006). A short proof of the Bulow-Klemperer auctions vs. negotiations result. *Economic Theory, 28*(2), 449–452.

Kreps, D. M. (1990). *A course in microeconomic theory*. Princeton, N.J.: Princeton University Press.

Lange, O. (1965). The computer and the market. In C. H. Feinstein (Ed.), *Socialism, capitalism and economic growth*. Cambridge University Press.

Lange, O., & Taylor, F. M. (1938). *On the economic theory of socialism*. Minneapolis: University of Minnesota Press.

Lerner, A. (1944). *The economics of control*. New York: Macmillan.

Mas-Colell, A., Whinston, M. D., & Green, J. R. (1995). *Microeconomic theory*. New York: Oxford University Press.

Moore, J. (1992). Implementation, contracts, and renegotiation in environments with complete information. In *Advances in Economic Theory* (Chap. 5, Vol. 5, pp. 182–282). Cambridge, U.K: Cambridge University Press.

Moore, J., & Repullo, R. (1988). Subgame perfect implementation. *Econometrica, 56*(5), 1191–1220.

Myerson, R. B. (1981). Optimal auction design. *Mathematics of Operations Research, 6*(1), 58–73.

Nisan, N. (2015). Algorithmic mechanism design: Through the lens of Multiunit auctions (Chap. 9). In H. P. Young & S. Zamir (Eds.), *Handbook of game theory with economic applications* (Vol. 4, pp. 477–515). Elsevier.

Palfrey, T. R. (2002). Implementation theory. In *Handbook of game theory with economic applications* (Vol. 3, pp. 2271–2326).

Roughgarden, T. (2016). *Twenty lectures on algorithmic game theory*. Cambridge University Press.

Russell, S. J., & Norvig, P. (2020). *Artificial intelligence : A modern approach* (SE - XVII, 1115 Seiten : Illustrationen, 4th edn). Boston: Pearson.

Sandholm, T. (1993). An implementation of the contract net protocol based on marginal cost calculations. In *Proceedings of the AAAI Conference on Artificial Intelligence*, 11.

Sandholm, T. (1998). Contract types for satisficing task allocation: I theoretical results. In *1998 AAAI Spring Symposium*.

Satterthwaite, M. A. (1975). Strategy-proofness and Arrow's conditions: Existence and correspondence theorems for voting procedures and social welfare functions. *Journal of Economic Theory, 10*(2), 187–217.

Shoham, Y., & Leyton-Brown, K. (2009). *Multiagent systems- algorithmic, game theoretic and logic foundation*. Cambridge University Press.

Smith, R. G. (1981). *A framework for distributed problem solving*. UMI Research Press.

Chapter 10
Mechanism Design and Auctions: Applications in OM

In Chap. 9, we discussed mechanism design and auctions. This chapter provides a discussion on applications of mechanism design and auctions.[1]

10.1 Sponsored Search Markets[2]

Advertising on the internet has reached another stratosphere since the beginning of this century. Unlike in the 1990s, advertisements are now handled by automated software systems. Online platforms generate value for advertisers by matching the content of the advertisement to the interests of the user on the platform. This is called *keyword-based advertising*, which was introduced by Google. An advertiser can pay either for an impression or an engagement done by the user. It is in the interest of the platform to match the advertisement's content with the user. This ensures the advertiser's return with future orders, generating revenue for the platform. The beauty of online advertising lies in its ability to match the advert's content with the user requirements in real time. This matching is done with the help of bots auctioning the space to the bid with the highest value (recall proxy bidder in Sect. 9.2.1). For example, advertisements shown to a user before the start of a YouTube video may depend on the video's content, the advertisement's content, and information about the user. Judging by the content of the video, YouTube bots hold auctions, receiving bids from the bots of the advertiser. Under the hood, YouTube scores the bids based on monetary value, relevance, user information, etc., to determine the winner of the auction. In this section, we discuss the auctions used in keyword-based advertising.

---

[1] It is important to note that the notations used in each section in this chapter are specific to that application.

[2] In this section, the OM actions are on the *process side* in the VCAP framework.

© The Author(s), under exclusive license to Springer Nature Singapore Pte Ltd. 2024
R. K. Amit, *Game Theory with Applications in Operations Management*, Springer Texts in Business and Economics, https://doi.org/10.1007/978-981-99-4833-8_10

211

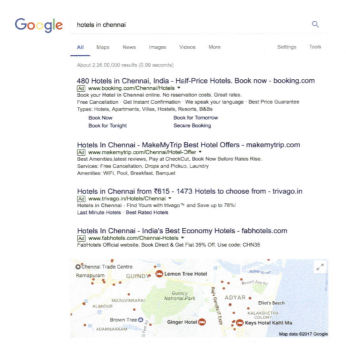

Fig. 10.1 Sponsored Results in Google Search

Search engines like Google reserve some spots on top of the search results for advertisers. These results are *sponsored*, indicated as (Ad) in Fig. 10.1, and the matching of these spots with the advertisers gives rise to *sponsored search markets*. Google's **AdWords** is the most well-known player in this space. But why auctions? Firstly, the number of keyword combinations is too big to imagine, and each combination can be a *search query*. It becomes impossible to set prices for each query. Secondly, the demand for a query and the advertisers willing to pay for that query fluctuates over time. Auctions allow search engines to match the *right* advertisement with a query to create value for users and advertisers. In Fig. 10.1, "hotels in chennai" is a search query. In this section, we provide details of the auction mechanism used by **AdWords**. Our discussion is motivated by Abrams et al. (2007), Bertsimas et al. (2016), and Haeringer (2018).

AdWords uses *pay-per-click* (PPC), which means that an advertiser pays only if its advertisement is clicked on by the user. Advertisers bid for different keyword combinations so that their advertisement can appear as the sponsored result of a query. As there are multiple slots available for sponsored results, **AdWords** sorts advertisements based on submitted bids and the quality score of advertisers. Furthermore, each advertisement has its *click-through rate* (CTR), which is computed using historical data. CTR is the percentage of clicks users make with respect to the

10.1 Sponsored Search Markets

Table 10.1 Click-through rate for different slots

Slot	Click-through rate (%)
Slot 1	10
Slot 2	8
Slot 3	5
Slot 4	1

number of times the advertisement is displayed. CTR is computed for each slot and reduces as one moves down the slots. Table 10.1 provides an illustration of CTRs for different slots.

10.1.1 Generalized Second-Price (GSP) Auction

Generalized second-price auction (GSP auction) follows the logic of second-price auctions discussed in Chap. 9. The steps in the GSP used by **AdWords** are as follows:

I. Final Score = Bid × Quality Score
II. Rank bidders according to the final score
III. The price (PPC) paid by the advertiser ranked k is the lowest price p such that $p \times$ Quality Score $>$ Final Score$_{k+1}$.

The GSP auction at **AdWords** is illustrated in Fig. 10.2. Assume there are three bidders for two slots "A" and "B" for query "hotels in chennai": Cleartrip, Makemytrip, and Yatra. Each bidder's bid and quality score is also shown.

AdWords combines each bidder's bid with its respective quality score to compute the final score, rank them based on the final score, and compute pay-per-click (PPC) for each bidder, as shown in Fig. 10.3. Slot A is assigned to Makemytrip's advertisement with a PPC of 3.01, and Slot B is assigned to Cleartrip's advertisement with a PPC of 1.01. Any bid $b_{MMT} > 3$ makes Makemytrip's final score higher than

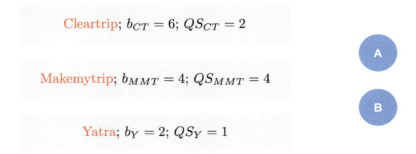

Fig. 10.2 GSP at AdWords

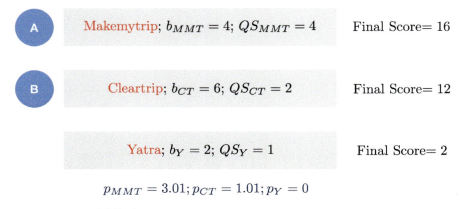

Fig. 10.3 GSP at AdWords—Determining Winners and PPC

Cleartrip's final score. If 0.01 is the minimum permissible bid increment, then PPC of Makemytrip is 3.01 to make it the winner of Slot A.

The GSP auction used by **AdWords** is in the class of efficient single-attribute auctions, and for such auctions, Myerson's lemma (Theorem 9.3) recommends a unique payment rule for truthfully implementing efficient allocation rule. The payment rule in the GSP auction is not the same as in Myerson's lemma, and hence, the GSP auction does not incentivize bidders to bid truthfully in dominant strategies.

Optimizing Delivery of Advertisements

In reality, operationalizing these auctions is complicated as multiple queries occur multiple times throughout the day. It is important for **AdWords** to optimize the delivery of advertisements to maximize its expected revenue under the budget constraints of the advertisers. This decision problem can be modeled as an optimization problem and is illustrated through an example.

Let us consider three queries that are used for searching for shoes: "Shoes", "Sports shoes", and "Trendy shoes". Each query's average frequency per day is in Table 10.2.

Table 10.2 Frequency of query

Query	Frequency
"Shoes"	30
"Sports shoes"	20
"Trendy shoes"	15

Four advertisers are bidding for these queries. Each advertiser decides on its daily budget and their bids for each search query, as shown in Table 10.3. As mentioned, the submitted bids may not be *truthful*.

AdWords assigns a quality score to each advertiser for each search query, as shown in Table 10.4.

10.1 Sponsored Search Markets

Table 10.3 Bids submitted for each query and daily budget

Query	Nike	H&M	Bata	HRX
"Shoes"	40	20	60	20
"Sports shoes"	80	20	10	60
"Trendy shoes"	20	60	30	10
Budget	120	80	90	70

Table 10.4 Quality scores

Query	Nike	H&M	Bata	HRX
"Shoes"	80	70	70	80
"Sports shoes"	90	10	60	80
"Trendy shoes"	30	90	60	20

Table 10.5 Final scores

Query	Nike	H&M	Bata	HRX
"Shoes"	3200	1400	4200	1600
"Sports shoes"	7200	200	600	4800
"Trendy shoes"	600	5400	1800	200

Table 10.6 Price-per-click (PPC) earned by the search engine

Query	Nike	H&M	Bata	HRX
"Shoes"	20.1	0.1	45.81	17.6
"Sports shoes"	53.43	0.1	3.43	7.6
"Trendy shoes"	6.76	20.1	10.1	0.1

Bids and quality scores are used to compute the final score of each bidder for each query, as shown in Table 10.5.

Using Step III of the GSP used by **AdWords**, PPC of each advertiser for each query can be computed. PPC values are shown in Table 10.6. The minimum permissible bid increment is 0.1.

Assume there are two advertisement slots available for each query. **AdWords** has to assign two advertisers to the slots available to maximize its expected revenue under the budget constraints of the advertisers. The order of appearance of the ads should not distort away from the order of the final score. For example, a possible combination of ads can be {Bata, HRX} but can never be {HRX, Nike}.

Since **AdWords** has only two slots, the possible search combinations for a search query are $\binom{4}{2} + \binom{4}{1} = 10$. A combination of advertisements in the slots is called a *slate*. Table 10.7 refers to all possible slates. A decision variable s_{ij} denotes the number of

216 10 Mechanism Design and Auctions: Applications in OM

Table 10.7 Possible slates

Query	Slate	Decision variable
"Shoes"	{Bata, Nike}	s_{10}
	{Bata, HRX}	s_{11}
	{Bata, H&M}	s_{12}
	{Nike, HRX}	s_{13}
	{Nike, H&M}	s_{14}
	{HRX, H&M}	s_{15}
	{Bata}	s_{16}
	{Nike}	s_{17}
	{HRX}	s_{18}
	{H&M}	s_{19}
"Sports shoes"	{Nike, HRX}	s_{20}
	{Nike, Bata}	s_{21}
	{Nike, H&M}	s_{22}
	{HRX, Bata}	s_{23}
	{HRX, H&M}	s_{24}
	{Bata, H&M}	s_{25}
	{Nike}	s_{26}
	{HRX}	s_{27}
	{Bata}	s_{28}
	{H&M}	s_{29}
"Trendy shoes"	{H&M, Bata}	s_{30}
	{H&M, Nike}	s_{31}
	{H&M, HRX}	s_{32}
	{Bata, Nike}	s_{33}
	{Bata, HRX}	s_{34}
	{Nike, HRX}	s_{35}
	{H&M}	s_{36}
	{Bata}	s_{37}
	{Nike}	s_{38}
	{HRX}	s_{39}

times a combination j appears for the search query i. The decision variables assigned for each combination are listed in Table 10.7.

Given the above information, **AdWords** can formulate an optimization problem to maximize its expected total revenue. The expected total revenue is $\sum_i \sum_j \pi_{ij}$, where π_{ij} is the expected revenue for slate ij. For slate 10, for s_{10} times, search query "Shoes" results in Bata at the first slot and Nike at the second slot. From Table 10.6, we know that PPC for Bata is 45.81, and for Nike, it is 20.1. The expected revenue π_{10} from slate 10 equals $\pi_{10} = s_{10} \times (PPC_1 \times CTR_1 + PPC_2 \times CTR_2) = s_{10} \times (45.81 \times 0.1 + 20.1 \times 0.08) = 6.189 s_{10}$.

10.1 Sponsored Search Markets

The expected payment of each bidder must satisfy the budget constraint. In Table 10.7, Bata is allocated the first slot in slates 10, 11, 12, 16, 25, 28, 33, 34, and the second slot in slates 21, 23, 30. The expected payment made by Bata is

$$0.1\Big[(45.81)(s_{10} + s_{11} + s_{12} + s_{16}) + (3.43)(s_{25} + s_{28}) + (10.1)(s_{33} + s_{34} + s_{37})\Big]$$

$+ 0.08\Big[(3.43)(s_{21} + s_{23}) + (10.1)(s_{30})\Big]$. Similarly, the expected payment of each advertiser can be computed.

Furthermore, **AdWords** should ensure that the total number of times the slates are shown for a given query does not exceed the frequency of that query. These constraints are called "query constraints". For example, $s_{10} + s_{11} + s_{12} + s_{13} + s_{14} + s_{15} + s_{16} + s_{17} + s_{18} + s_{19} \leq 30$ is the query constraint for query "Shoes", which has frequency 30 (refer Table 10.2).

The decision variables are *integer-valued* s_{ij}. Solving the optimization problem in Excel solver, the optimal values of decision variables are $s_{11} = 19, s_{13} = 9, s_{15} = 2, s_{20} = 19, s_{23} = 1, s_{30} = 3, s_{31} = 12$.

10.1.2 *Vickrey-Clarke-Groves (VCG) Mechanism*

We discussed Vickrey-Clarke-Groves (VCG) mechanism in Sect. 9.3.4, and proved that VCG payment rule in Eq. ★ *implements* the efficient allocation rule in dominant strategies.

VCG mechanism is used by **Facebook** for its display advertisements. The nature of these advertisements is quite different, and the market designed for the purpose reflects that. Unlike the sponsored search, which uses a keyword, there is no keyword, which leaves the advertisers with limited information regarding the user's requirements. The only information available is the user's cookies. In this situation, it becomes hard for the advertiser to gauge the valuation of a display slot properly. Each display slot has a different CTR, making it hard for advertisers to pick a bidding strategy. For this reason, **Facebook** reduces this problem of advertisers by using the VCG mechanism that elicits truthful bidding behavior in dominant strategies (recall quote by John Hegeman in Sect. 9.3.4). This reduces the advertisers' burden of computing the optimal bidding strategy and thus allows them to focus on the quality of the advertisement, which in turn helps **Facebook** as it improves the user experience.

We use the notation of Sect. 9.3.4. As VCG induces truthful bidding, the efficient allocation rule $\mathbf{x}(\mathbf{b})$ can be written as $\mathbf{x}(\mathbf{v})$. $\mathbf{x}(\mathbf{v})$ maximizes $\sum_{i=1}^{n} v_i(S_i)$. S_i is the slot assigned to advertiser i. The payment rule for advertiser i, which captures the valuation loss to other advertisers can be written as

$$p_i(\mathbf{v}) = \sum_{j \neq i} v_j(S'_j) - \sum_{j \neq i} v_j(S_j)$$

where S_j is the slot assigned to advertiser j using the efficient allocation rule $\mathbf{x}(\mathbf{v})$ that maximized $\sum_{i=1}^{n} v_i(S_i)$. S'_j is the slot assigned to advertiser j using the efficient

218   10 Mechanism Design and Auctions: Applications in OM

Table 10.8 Valuation of advertisers for cookie based on "Sports Shoes"

Nike	H&M	Bata	HRX
80	20	10	60

Table 10.9 Advertiser's valuations of slots for cookie based on "Sports Shoes"

Advertiser	Slot 1	Slot 2	Slot 3	Slot 4
Nike	8	6.4	4	0.8
HRX	6	4.8	3	0.6
H&M	2	1.6	1	0.2
Bata	1	0.8	0.5	0.1

allocation rule $\mathbf{x}'(\mathbf{v})$ that maximized $\sum_{j \neq i} v_j(S'_j)$. In other words, S'_j is the slot assigned to advertiser j using the efficient allocation rule $\mathbf{x}'(\mathbf{v})$ when advertiser i is not present in the mechanism.

Let us illustrate the VCG mechanism for **Facebook** through an example that uses a cookie based on "Sports Shoes". As earlier, there are four advertisers who bid truthfully in the VCG mechanism, with valuations are shown in Table 10.8.

There are four display slots with click-through rate (CTR) shown in Table 10.1. Using CTR, each advertiser's valuation of each slot is shown in Table 10.9.

Based on vlautions, the order of advertisements is Nike, HRX, H&M, and Bata. Since Bata is in Slot 4, it does not cause harm to anyone and its payment is thus zero. But this is not the case for other advertisers. H&M's presence causes Bata to move to Slot 4, where its payoff is 0.4. Without H&M, Bata would have been in Slot 3 with valuation 0.5. Hence, H&M's payment $= (0.5 - 0.1) = 0.4$. Similarly, HRX's presence causes H&M to move to Slot 3 and Bata to Slot 4. The payments of HRX and Nike are

$$\text{HRX's payment} = (1.6 - 1) + (0.5 - 0.1) = 1$$
$$\text{Nike's payment} = (6 - 4.8) + (1.6 - 1) + (0.5 - 0.1) = 2.2$$

The total revenue of **Facebook** is $0 + 0.4 + 1 + 2.2 = 3.6$.

10.2 Community Sensing[3]

Community sensing entails attaining information about a system (or an environment) by aggregating information reported by multiple independent agents. Community sensing aims to accurately measure the true state of the system. This is typically done

---

[3] In this section, the OM actions are on the *process side* in the VCAP framework.

10.2 Community Sensing

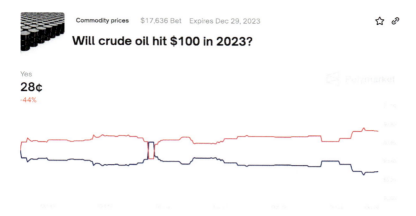

Fig. 10.4 Prediction Market. *Source* https://polymarket.com/

by incentivizing the agents who report truthfully. One example of community sensing is *prediction markets*. In prediction markets, the agents place a bet on the outcome that they perceive may occur, and are rewarded in case of accurate predictions. These markets work on a mechanism along the principles of a *double-sided auction*, and the market eventually settles on a probability reflecting the agents' consensus. Prediction markets are used for scenarios for which the outcome can be objectively verified ex-post. Figure 10.4 illustrates the use of prediction markets to gather information about the price of crude oil at the end of 2023 at *Polymarket*. The logic behind such markets is simple: independent and rational agents use everything in their power to obtain as much information about the event to secure the highest rewards from the market. Prediction markets provide a good proxy for a particular event's probability. In this example, the probability for crude oil to cross $100 per barrel is 0.28.

For outcomesCommunity sensinghat cannot be verified *ex-post*, peer prediction systems are used. The aggregated value reported by the agents is a proxy for the true state of the system—instead of comparing the prediction of an agent with the *ex-post* outcome, the prediction is compared with this aggregated value. Scoring rules that reward an agent for truthfully reporting their values are devised.[4] Deviations from the aggregated value are penalized. As discussed in Chap. 9 for truth-telling mechanisms, the scoring rules should satisfy *individual rationality*, *incentive compatibility*, and *computational tractability* (Chen & Pennock, 2010). It can be inferred that a community sensing problem is a mechanism design problem, and scoring rules are similar to payment rules.

We discuss community sensing for air pollution estimation, which is motivated by Faltings et al. (2013). Traditional measurement of air pollution requires large and expensive installations. With advances in micro-sensors that can be used for

---

[4] One example of misreporting in community sensing is fake product reviews. Refer "A new study analyses the murky world of fake Amazon reviews" (The Economist; September 3, 2020).

Fig. 10.5 Micro-sensors for Air Pollution

individual agents for estimating air pollution (Fig. 10.5), community sensing can be used for getting a detailed picture of air pollution.

Community sensing architecture is shown in Fig. 10.6. Agents observe the ground reality and submit their observations to the central planner, who aggregates the information to obtain a good estimate of the ground reality. This poses an issue of quality control, as the center has no control over the quality of the measurements it receives because the agents need to exert considerable effort in observing the state of the system and reporting their observations accurately. For example, an agent may have to take several readings over a short time period to obtain an accurate picture of the ground reality. How can the agents be rewarded for truthfully reporting their observations?

We begin the discussion by deriving two of the most common scoring rules—*logarithmic* and *quadratic scoring rules*. Let us assume that air pollution has two possible states {low, high}, where "low" can occur with probability p and "high" can occur with probability $1 - p$. Scoring rule $S(x, p)$ pays $f(x)$ to agent i if he reports $p^i = x$ as the probability of "low" state to the central planner. The expected payment $\pi^i(x, p)$ of agent i using scoring rule $S(x, p)$ is given in Eq. 10.1.

$$\pi^i(x, p) = pf(x) + (1 - p)f(1 - x) \qquad (10.1)$$

For truthful reporting, scoring rule $S(x, p)$ should have the payment functions $f(x)$ such that $\pi^i(x, p)$ is maximized at $p^i = p$. This can be achieved by differentiating $\pi^i(x, p)$ with respect to x to get the first-order condition $pf'(x) = (1 - p)f'(1 - x)$.

10.2 Community Sensing

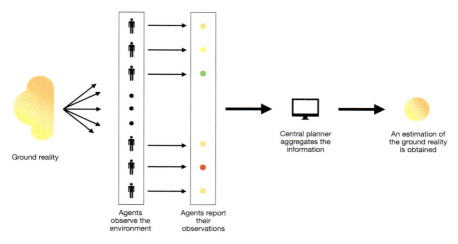

Fig. 10.6 Community Sensing Architecture

One possible solution that satisfies the condition is $f'(p) = 1/p$, with payment function $f_L(x) = \log x$, and the scoring rule is called *logarithmic scoring rule*. The other possible solution is $f'(p) = (1 - p)$, with payment function $f_Q(x) = x - \frac{x^2}{2}$, and the scoring rule is called *quadratic scoring rule*.

Let us consider community sensing for air pollution, where air pollution has two possible states {low, high}, where "low" can occur with probability p and "high" can occur with probability $1 - p$. Multiple agents report the values of pollution observed, and the ground reality is estimated by aggregating all reports. The estimates indicate $p = 0.2$ (the probability of low pollution) and $1 - p = 0.8$ (the probability of high pollution. Agents {A, B} are among the agents who reported their probabilities $(p^A$ and $p^B)$. Agent A reports truthfully, while Agent B manipulates the observed probability, as shown in Fig. 10.7. In this case, Agent B may be a large polluter and strategically report false information to paint a very different picture from reality.

If the central planner uses the logarithmic scoring rule with payment function $f_L(x) = \log x$, then the expected payment to Agent A is

$$\pi^A(x, p) = p \log x + (1 - p) \log(1 - x)$$
$$= 0.2 \log(0.2) + (0.8) \log(0.8)$$
$$= -0.217$$

Similarly, the expected payment to Agent B is

$$\pi^B(x, p) = p \log x + (1 - p) \log(1 - x)$$
$$= 0.2 \log(0.6) + (0.8) \log(0.4)$$
$$= -0.362$$

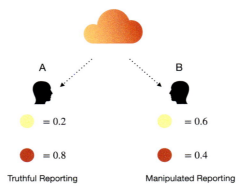

Fig. 10.7 Truthful and Manipulated Reporting

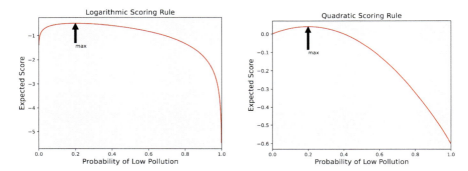

Fig. 10.8 Comparing Scoring Rules

Using the logarithmic scoring rule, Agent A obtains higher payment than Agent B. The quadratic scoring rule can also be used. The plots shown in Fig. 10.8 compare the expected payoff to an agent using the logarithmic and quadratic scoring rules. The quadratic scoring rule penalizes an agent much more sharply for deviations from the truth.

10.3 Physical Internet[5]

The conventional methods employed for transporting physical goods globally face economic, environmental, and social challenges. The presence of economic inefficiency can be attributed to the inadequate utilization of transportation and storage capacities. A carrier, whether it be a truck, cargo flight, or liner, often operates at a significantly lower capacity than what is permitted. There is a multitude of evidence

---

[5] In this section, the OM actions are on the *process side* in the VCAP framework.

10.3 Physical Internet

available to substantiate this inefficiency. As a result of the extensive utilization of trucks, freighters, and containers, fuel consumption is heightened, thereby contributing to an overall increase in energy expenditure. Moreover, the carbon dioxide (CO_2) emissions generated by carriers pose a significant challenge in the global efforts to mitigate climate change. Finally, when considering the societal aspect, it can be observed that the overall quality of life for individuals working in logistics and transportation has deteriorated due to the prolonged periods they must spend engaged in the task of transporting shipments.

In order to provide additional evidence of the inefficiencies, Montreuil (2011) presents a comprehensive compilation of 13 indicators of non-sustainability, supported by empirical data from real-world sources. One can argue that the inefficiencies primarily stem from the incremental rise in demand resulting from globalization. On the opposite end of the spectrum, there exist brick-and-mortar industries that solely maintain inventory within their physical establishments, lacking the capacity to identify appropriate clientele. Nevertheless, the fulfillment of demands and the provision of services often come at the expense of inefficiencies.

One of the limited strategies available to mitigate these inefficiencies involves the establishment of horizontal collaboration. Collaboration serves to mitigate uncertainty and enhance margins. However, carriers are reluctant to collaborate due to concerns about losing competitive advantage. Simultaneously addressing these various efficiencies necessitates the application of innovative approaches, resulting in the emergence of a novel conceptual framework known as *Physical Internet* (PI). PI plays a crucial role in enhancing the efficiency of freight transportation. Figure 10.9 presents the vision of PI.

The information and telecommunications community faced significant challenges due to its limited capacity to exchange information within private networks. In order to facilitate efficient and smooth sharing, they drew inspiration from the logistics industry by establishing the *Internet* as the information superhighway. The advent of the internet has facilitated the development of the World Wide Web and enabled

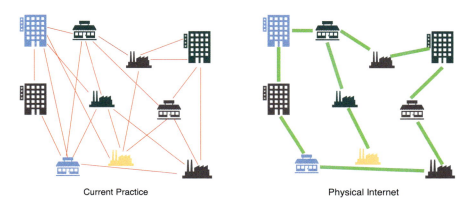

Fig. 10.9 Vision of Physical Internet. Adapted from Ciprés and de la Cruz (2019)

digital mobility. Data is organized into packets and transmitted in a standardized manner based on the Transmission Control Protocol (TCP) or Internet Protocol (IP). The advent of standardized transmission protocols facilitated the development of the digital internet, and it is precisely this principle that the Physical Internet seeks to emulate. Several aspects of the analogy between physical and digital internet are briefly highlighted in the following points.

- **Encapsulation**: Digital internet transfers packets that contain embedded information. The packets are encapsulated to include headers containing information about their identities and routing destinations. PI uses smart, green, world-standard, and modular containers to ease the handling, storage, and transporting activities. These containers are commonly referred to as π-**containers**.
- **Universal connectivity**: Without further elaboration, the digital internet facilitates global connectivity. In a similar vein, it is imperative for PI to possess universal connectivity in order to effectively minimize setup times at interconnections. One of the primary objectives of connectivity is to minimize the expenses and time associated with loading and unloading activities at interconnections.
- **Embedding**: Just like mandatory access control in the digital internet, π-containers possess distinctive global identifiers. Smart tags are used to establish a connection between the container and its agent to enhance transportation transparency.
- **Multi-segment intermodality**: In the realm of digital networking, the transmission of data packets does not occur in a direct manner from the source to the destination. Data packets are transmitted via routers and cables. The authoritative information within the packet can originate from multiple discrete routers. However, prior to being delivered, they are packaged together. Similarly, it is advisable for the logistics transportation system that is currently dominated by point-to-point operations to transition towards a multi-segment approach. The process of fragmentation necessitates the involvement and dedication of numerous individuals yet ultimately alleviates the burden placed upon them.

In addition to these factors, the successful implementation of PI necessitates a multi-tier conceptual framework and the establishment of an open global supply network. A crucial aspect involves the design of products with the aim of minimizing spoilage in π-containers, which consequently leads to a decrease in the frequency of less than truckload movements. In summary, PI aims to recreate the digital internet by establishing standardized logistics transportation equipment and requiring connectivity from numerous independent agents. For further understanding of the state of the art of PI, we refer to Sternberg and Norrman (2017), Treiblmaier et al. (2020), and Chen et al. (2022).

The potential for game theory in the domain of the physical internet is readily apparent. This section focuses on two primary domains: coordinating agents at interconnections to facilitate efficient logistical services and implementing incentives to motivate agents towards achieving global optima. While some scholarly articles emphasize the significance of adopting a game-theoretic perspective in the field of PI, it is important to note that research in this area is still in its early stages. For example, the study conducted by Plasch et al. (2021) examines the necessity of collaboration and the underlying motivations and factors contributing to its success. Niu et al. (2022) ascertain the circumstances under which both service providers derive advantages from the implementation of PI, taking into account different levels of logistics efficiency and service competition.

To make the physical internet successful, the *right carriers* should carry the *right packages* on the *right routes*. This scenario ensures a mutually beneficial outcome for both carriers and shippers. The process of allocating shipments to carriers is of utmost importance. The alignment of packages with the offerings in PI hubs should be dynamic and localized due to the presence of interconnections. Furthermore, in the context of transportation, carriers should be granted the privilege of obtaining transportation rights to operate inside a specific route. Auction theory can be used to achieve these goals. However, single-attribute auctions, discussed in Sect. 9.3, have limited applicability in PI due to the presence of complementarities and substitutes in PI. *Combinatorial auctions* (refer Sect. 9.3.4) mitigate the problem of complementarities and substitutes and can be used for allocating requests to the carriers (Pan et al., 2014), and our discussion is based on Pan et al. (2014).

The auction mechanism proposed is *first-price sealed-bid combinatorial auction*, where shippers act as auctioneers and carriers act as bidders at each interconnection point in the PI network. To guarantee global efficiency in the PI network, three rules are needed

Reallocation and co-delivery are allowed: In each PI hub, requests from shippers are auctioned and allocated to the best possible carrier. It is expected that a request can be carried by carrier A on one edge, followed by carrier B on the other edge of the network.

No halfway dropping: This rule serves as a bound for the previous rule. Each carrier has to submit bids for a request such that its entire transportation from source to destination is feasible. For instance, in the previous rule, if carrier B is not in the picture, then carrier A has to operate in the subsequent edge.

Non-arbitrary carrier shifting: If more than two carriers offer the best price, then the request should go to either one of the carriers. No extra amount should be charged by the chosen carrier.

The sequence of the auction mechanism is as follows: At each PI hub, carriers are required to handle an exponential number of bundles derived from the requests in order to bid based on their preferences. Shippers are required to undertake the processing of the bids that have been submitted, a task that is characterized by its demanding nature in terms of both time and effort. Thus, both shippers and carriers employ *proxy agents* (recall proxy bidder used while discussing the revelation principle in Sect. 9.2.1) to assist them. The results generated by both proxies are transmitted to the winner determination model. The results generated by the winner determination model are disseminated to both shippers and carriers to conclude the auction process. The detailed discussion of the auction mechanism is as follows:

Proxy-bidding Agents

Proxy-bidding agents assist carriers in bidding according to their preferences. The objective of these agents is to return the best possible bids given the private information and preferences of the corresponding carriers. It is attained in two steps: determining feasible request bundles and proposing the bid for those bundles.

Determining feasible request bundles: A request bundle RB is a subset from a given pool of requests R. Bundle RB is feasible for a carrier if it satisfies route compatibility and capacity of the carrier. Using the "no halfway dropping" rule, each request has a predetermined route, and a request is *compatible* with route i if the request can be completed by the carrier's truck traveling on route i. All compatible requests on route i are collated as G_i, and all such compatibles are grouped together as $G = \{G_1, G_2, \ldots, G_n\}$. In addition, the capacity limitations of the carrier further narrow down to produce feasible request bundles. Precisely, $RB_1, RB_2, \ldots, RB_K$ are feasible request bundles if $RB_k \subseteq G_i$, where $k \in \{1, \ldots, K\}$, $i \in \{1, \ldots, n\}$, and $\sum_{r \in RB_k} v_r \leq CP_t$, where v_r is the volume of request r and CP_t is the capacity of the truck of carrier t. It is assumed that each carrier operates a single truck, and $t \in \{1, \ldots, T\}$.

Proposing bids for feasible request bundles: The bid function for a carrier is based on the transportation cost $CT_{RB_k}^t$ incurred by request bundle RB_k and preferred margin $mr_{RB_k}^t$ of the carrier. Relatively, the bid price for a request bundle $P_{RB_k}^t = CT_{RB_k}^t \times (1 + mr_{RB_k}^t)$. $P_{RB_k}^t$ is a naive way to calculate bids, which stands as a future research direction.

Proxy-auctioning Agents

In Sect. 9.3.1, we mentioned that an auction mechanism has allocation and payment rules. Proxy-auctioning agents stand in place of shippers to allocate requests and determine their payments. In other words, proxy-auctioning agents break down the bid for the request bundle to payment for each request. Payment PS_r^t of request r is calculated on pro-rata basis using distance d_r and volume v_r of request r, is given in Eq. 10.2.

$$PS_r^t = P_{RB_k}^t \times \frac{d_r v_r}{\sum_{r \in RB_k} d_r v_r} \tag{10.2}$$

In this case, truthful bidding is not guaranteed, as the payment rule is not the VCG payment rule (refer Theorem 9.4). In addition to PS_r^t, proxy agents derive *payment rate* $RS_r^t = \frac{PS_r^t}{d_r v_r} = \frac{P_{RB_k}^t}{\sum_{r \in RB_k} d_r v_r}$, which can be inferred as *logistic efficiency*. A high payment rate indicates low logistic efficiency. By definition, RS_r^{ti} ensures that all the requests in a bundle have the same efficiency. This concept is important in the case of *reallocation*—reallocation is permitted only when lower payment rate is available.

Winner Determination Problem (WDP)

In Sect. 9.3.4, we mentioned that the winner determination problem (WDP) belong to the class of set packing problems, hence **NP-Complete**. WDP is formulated as an integer linear programming problem for PI settings as given in Eq. 10.3. In the formulation, we preclude reallocations.

$$\min \sum_t \sum_{RB_k} PB^t_{RB_k} x^t_{RB_k}$$

$$\text{subject to } \sum_t \sum_{\substack{RB_k \\ r \in RB_k}} x^t_{RB_k} = 1 \quad \forall r \in RB_k$$

$$\sum_{RB_k} x^t_{RB_k} \leq 1 \quad \forall t$$

$$x^t_{RB_k} \in \{0, 1\} \quad \forall t, RB_k$$

(10.3)

In the formulation, the objective function is to determine the winners so the total cost of shipping all the requests is minimized. The first constraint ensures that each request is allocated only once. The second constraint ensures that each carrier wins only one bundle. $x^t_{RB_k}$ is a binary decision variables with $x^t_{RB_k} = 1$ when bundle RB_k is allocated to carrier t. The case presented there is simple, yet illustrates the potential of combinatorial auctions in the PI network.

The literature of game-theoretic modeling of PI is still limited. However, there are significant research opportunities in that direction (Kong et al., 2018). The exploration of novel mechanisms inspired by the physical internet, such as the optimal bid function, addressing information asymmetry among bidders, and mitigating cartel formation to promote socially desirable outcomes, is a captivating and intricate study area. The development of intelligent bots as proxies, the utilization of blockchains to facilitate the coordination of bidding processes and the creation of platforms that enable seamless bidding are intriguing areas of research that enhance the applicability of the physical internet to modern transportation.

10.4 Mechanism Design for Systems Engineering[6]

Systems engineering is a holistic approach that involves conceptualizing, integrating, and administrating complex systems, including several components, their interrelationships, and ultimate goals. The process comprises a range of steps, which include requirements analysis, design, integration, verification, validation, risk management, and lifecycle planning. The main objective of this procedure is to guarantee the efficient integration and alignment of diverse systems, enabling them to operate harmoniously and accomplish their intended objectives during their entire duration.

The design framework should possess characteristics that promote sustainability, robustness, consistency, and rationality within the system. Unfortunately, Hazelrigg (2012) asserts that the assurance of a rational design is contingent upon the presence

---

[6] In this section, the OM actions are on the *process side* in the VCAP framework.

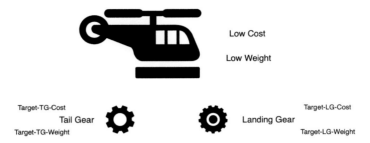

Fig. 10.10 Designing Gears for Light Combat Helicopter

of a single decision-maker. Given the involvement of several decision-makers in each aspect of the system, it is evident that their respective sets of alternatives, beliefs, preferences, and objectives exhibit significant variations. This is illustrated through an example motivated by Hazelrigg (2012).

A new combat helicopter[7] is designed with *low cost* and *low weight* goals. A team of engineers is working on designing a landing gear with target cost (Target-LG-Cost) and target weight (Target-LG-Weight), as shown in Fig. 10.10. Another team of engineers is independently working on designing a tail gearbox with target cost (Target-TG-Cost) and target weight (Target-TG-Weight).

The team working on the landing gear converges on initial design "D" of the landing gear with cost (Cost-LG-D) exceeding target cost (Target-LG-Cost), with weight (Weight-LG-D) within the target, while the team working on the tail gear converges on initial design "D" of the tail gear with weight (Weight-TG-D) exceeding target cost (Target-TG-Weight), with cost (Cost-TG-D) within the target. The LG team revises the design to "DD" with lighter material (Titanium) but more expensive than steel. Similarly, the TG team revises the design with a cheaper material than steel but heavier than steel. Though individual teams achieve their respective goals, the new design (DD) of combat helicopters is both costly and heavier than the best design. This discussion is shown in Fig. 10.11. Each team's performance is measured based on their respective targets; the incentives are not *right* for converging to the best design.

The situation can be modeled as a mechanism design problem, as shown in Fig. 10.12. The new combat helicopter's design head (or mechanism designer) can ask individual teams to communicate the design parameters of their respective components that can be used to design the helicopter. If the LG team communicates its parameter as θ_1, while the TG team communicates its parameter as θ_2, the design head converges on the design with parameter $\theta = (\theta_1, \theta_2)$. As discussed previously, the teams can act strategically and may not communicate the best design parameters. The design head can design a mechanism to incentivize the teams to *truthfully* communicate the parameters so that the design head can achieve (or implement) the best design for the helicopter.

---

[7] An example is Light Combat Helicopter (LCH) *Prachand* in India.

Problems

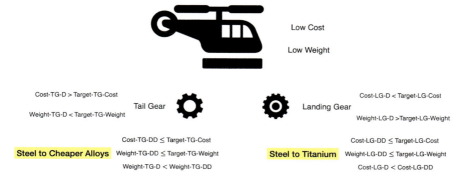

Fig. 10.11 Incentives Issues in Designing Gears for Light Combat Helicopter

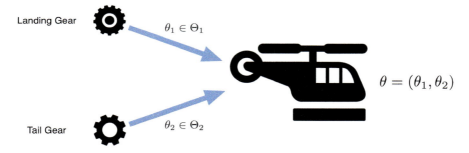

Fig. 10.12 Mechanism Design Framework in Designing Gears

This provides the relevance of mechanism design in the field of systems engineering and its capacity to generate tangible benefits. However, it is worth noting that mechanism design has not yet been widely applied in the field of systems engineering despite its potential for addressing various challenges in this domain.

Problems

Problem 10.1 Let there be a sponsored search market for two similar keyword queries "Pizza near me" and "Restaurants nearby". Three local businesses, Pizza Junction, Food Hub, and The Dream Kitchen, bid for two sponsored slots. The relevant information is provided in the Tables 10.10, 10.11, 10.12 and 10.13.
Find the price-per-click paid to the search engine by each advertiser for each query. How should the search engine allocate the slots to the advertisements? ◀

230   10   Mechanism Design and Auctions: Applications in OM

Table 10.10 Click-through-rate for different slots

Spot	Click-through-rate (%)
First	10
Second	5

Table 10.11 Bids submitted for each query and daily budget of an advertiser

Query	Pizza junction	Food hub	The dream kitchen
"Pizza near me"	12	10	4
"Restaurants nearby"	6	9	14
Budget	15	18	17

Table 10.12 No. of hits for each search query

Query	Hits
"Pizza near me"	20
"Restaurants nearby"	50

Table 10.13 Quality score of each bidder for different search queries

Query	Pizza junction	Food hub	The dream kitchen
"Pizza near me"	9	8	4
"Restaurants nearby"	6	8	9

Problem 10.2 A governmental agency wishes to quantify the levels of pollution in a city by placing a probability distribution over three possibilities: low (l), medium (m), and high (h). Based on the time of the year and activities of the previous week, there is a public prior about the pollution level. This prior states the probabilities as ($l = 0.3, m = 0.6, h = 0.1$). After aggregating all reports, the pollution level is estimated to be ($l = 0.2, m = 0.5, h = 0.3$). The agent incurs a cost of $c = 1$ unit for calibrating his instrument to obtain a truthful observation. Let the payment function be $a + b \cdot S(x, p)$, where $S(x, p)$ is the logarithmic scoring rule, and b is a scaling constant.

   i What is the minimum value of b for which the agent calibrates his instrument? The agent reports his observation as the public prior otherwise.
   ii How does the answer change if the quadratic scoring rule is used?   ◀

Problem 10.3 Consider a scenario where Company A produces 20 units of a product and needs to transport them to Company B. Compare the costs involved in two different supply chain scenarios: the *traditional supply chain*, where Company A ships the products directly to Company B, and *Physical Internet*, where a consolidation center and distribution hub are introduced. In the traditional supply chain, the transportation cost is $1 per unit, and Company A bears the shipping cost. In Physical Internet, Company A sends the products to a consolidation center, incurring a handling and consolidation fee of $10. The consolidated shipment is then transported to a distribution hub, with a transportation cost of $0.5 per unit. The distribution hub delivers the products to Company B. ◄

References

Abrams, Z., Mendelevitch, O., & Tomlin, J. (2007). Optimal delivery of sponsored search advertisements subject to budget constraints. In *Proceedings of the 8th ACM Conference on Electronic Commerce, EC 2007* (pp. 272–278). Association for Computing Machinery.

Bertsimas, D., Allison, K., & Pulleyblank, W. R. (2016). *The analytics edge*. Dynamic Ideas LLC Belmont.

Chen, Y., & Pennock, D. M. (2010). Designing markets for prediction. *AI Magazine, 31*(4), 42–52.

Chen, S., Su, L., & Cheng, X. (2022). Physical internet deployment in industry: Literature review and research opportunities. *Industrial Management and Data Systems, 122*(6), 1522–1540.

Ciprés, C., & de la Cruz, M. T. (2019). The physical internet from shippers perspective. In *Towards user-centric transport in Europe: Challenges, solutions and collaborations* (pp. 203–221).

Faltings, B., Li, J. J., & Jurca, R. (2013). Incentive mechanisms for community sensing. *IEEE Transactions on Computers, 63*(1), 115–128.

Haeringer, G. (2018). *Market design : Auctions and matching*. MIT Press.

Hazelrigg, G. A. (2012). *Fundamentals of decision making for engineering design and systems engineering*. George A.

Kong, X., Huang, G. Q., Luo, H., & Yen, B. P. (2018). Physical-internet-enabled auction logistics in perishable supply chain trading: State-of-the-art and research opportunities. *Industrial Management and Data Systems, 118*(8), 1671–1694.

Montreuil, B. (2011). Toward a physical internet: meeting the global logistics sustainability grand challenge. *Logistics Research, 3*, 71–87.

Niu, B., Dai, Z., Liu, Y., & Jin, Y. (2022). The role of physical internet in building trackable and sustainable logistics service supply chains: A game analysis. *International Journal of Production Economics, 247*, 108438.

Pan, S., Xu, X., Ballot, E., et al. (2014). Auction based transport services allocation in physical internet: A simulation framework. In *5th international conference on information systems, logistics and supply chain*.

Plasch, M., Pfoser, S., Gerschberger, M., Gattringer, R., & Schauer, O. (2021). Why collaborate in a physical internet network?—motives and success factors. *Journal of Business Logistics, 42*(1), 120–143.

Sternberg, H., & Norrman, A. (2017). The physical internet-review, analysis and future research agenda. *International Journal of Physical Distribution and Logistics Management, 47*(8), 736–762.

Treiblmaier, H., Mirkovski, K., Lowry, P. B., & Zacharia, Z. G. (2020). The physical internet as a new supply chain paradigm: a systematic literature review and a comprehensive framework. *The International Journal of Logistics Management, 31*(2), 239–287.

Index

A

Advance Market Commitment (AMC), 110
Airline alliance game, 68
Airline alliances, 65
Artificial intelligence, 159
  explainability, 159
  interpretability, 159
Asymmetric auctions, 204
Auctions, 193
  combinatorial auctions, 194
  desiderata, 193
  Dutch auctions, 193
  efficient auctions, 194
  English auctions, 193
  first-price sealed-bid (FPSB), 23, 51, 184
  generalized second-price (GSP) auction, 213
  Jain auctions, 184
  Japanese auctions, 193
  knapsack auctions, 195
  multiple attributes, 193
  optimal auctions, 194, 199
    virtual valuation, 201
  second-price sealed-bid (SPSB), 33, 51, 184
  single attribute, 193
Aumann–Shapley pricing, 164
  axioms, 165

B

Babbling equilibrium, 117
Backward induction, 88
Balanced collection, 132
Balanced game, 132

Bayesian games, 51
Bayesian-Nash equilibrium, 48, 51
  in mechanism design, 173
Bayes' rule, 100
Behavior strategy in extensive-form games, 93
Best response strategy, 34
Bhagavad Gita, 3
Bi-matrix games, 12
Bitcoin, 78
Bitcoin mining, 80
Bitcoin mining game, 80
Blockchains, 44, 78
Bondareva–Shapley theorem, 132
Booking limits, 66, 67
Bounded set, 37
Braess's paradox, 35, 61, 64
Buffer, 4
  capacity, 4
  inventory, 4
  time, 4
Bullwhip effect, 7
Business platforms, 179
Byzantine Generals Problem, 78

C

Capacity decisions, 105
  fungible capacity, 105
  specialized capacity, 105
Cardinality, 128
Centipede game, 97
Cheap talk, 14, 44, 78, 107
  in operations management, 115
Church-Rosser property, 33

© The Editor(s) (if applicable) and The Author(s), under exclusive license to Springer Nature Singapore Pte Ltd. 2024
R. K. Amit, *Game Theory with Applications in Operations Management*, Springer Texts in Business and Economics, https://doi.org/10.1007/978-981-99-4833-8

234   Index

Closed set, 37
Coalitional games, 12
  with transferable utility, 17
Code sharing, 66
Combinatorial auctions, 205
  direct mechanisms, 205
  in physical internet, 225
  indirect mechanisms, 206
Common knowledge, 23
Community sensing
  logarithmic scoring rule, 220
  quadratic scoring rule, 220
Complementarity, 128
Concave game, 149
Constant-sum games, 128
Contract net, 182
Conventions, 45
Convex games, 128
Convex set, 37
Cooperative games, 12
Coordination theory, 7
Core, 12, 125, 130
  in inventory games, 147
  in service systems, 149
  of trade game, 131
Correlated equilibrium, 44, 45, 48
Correspondence, 37
Cournot duopoly, 23, 58, 87
COVID-19 vaccines, 106
Critical fractile, 7, 72
Cryptographic hash function, 78
  collision resistance, 78
  hiding, 78
  puzzle-friendliness, 78
Cryptographic hash puzzles, 78, 80

D
Dagen-H, 45
Decentralized ledgers, 78
Delayed differentiation, 7
Detail-free
  in mechanism design, 199
  in supply chain contracts, 77
Digital signatures, 78, 79
Discount factor, 88
Diseconomies of scale, 107
Dominant strategy, 32
Dominant strategy equilibrium, 32
  in mechanism design, 178
Dominated strategy, 33
Domination
  Pareto domination, 31

strict domination, 32
strict Pareto domination, 31
weak domination, 32
Double marginalization, 7, 69, 71

E
Economic order quantity (EOQ), 5
Economies of scale, 148
Economies of scope, 148
Efficient payoff, 129
Electronic data interchange (EDI), 8
Entry deterrence game, 85
Equilibrium selection problem, 35

F
Farsighted stability, 152
Feasible payoff, 129
Finite horizon bargaining game, 87
  alternating offer bargaining protocol, 87
Flow games, 156

G
Games
  in characteristic form, 125
  in extensive form, 85
  in normal form, 21
Games with perfect recall, 93
Gibbard–Satterthwaite theorem, 181
Gig-economy Game, 46
Gilder's law, 78
Global Alliance for Vaccines and Immuniza-
  tion (GAVI), 110
Golden ratio, 93
Great divergence, 1
GREEN game, 154

H
H-dagurinn, 45

I
Imputation, 129
Incentive compatibility, 175, 190
Individually rational payoff, 129
Individual rationality, 189
Inessential and essential games, 128
Information partition, 48
Information sets, 16
Internet-of-Things (IoT), 46
Inventory games

Index

characteristic-form representation, 143
extensive-form representation, 121
normal-form representation, 57

J
Job market signaling, 116

K
Kakutani fixed-point theorem, 37
Kingman's formula, 5, 148
Kuhn's theorem, 93

L
Leadership in supply chains, 112
Lean production systems, 1
Lemke-Howson algorithm, 38, 42
Linear complementarity problem (LCP), 38

Linear feasibility problem, 47
Linear production game, 127
Linear programming, 28
  complementary slackness condition, 26, 29
  simplex algorithm, 28
  strong duality theorem, 28
  weak duality lemma, 133
Little's law, 4, 148
Logistics networks, 155

M
Mahatma Gandhi's talisman, 30
Markov-Perfect Bayesian Nash Equilibrium, 120
  action rule, 120
  signaling rule, 120
Matching pennies, 22
  mixed extension, 24
Matrix games, 25
Mechanism design, 32, 171
  algorithmic mechanism design, 195
  formalism, 185
  implementation, 186
    strong, 186
    weak, 181, 186
  in systems engineering, 227
Metcalfe's law, 78
Minimax theorem, 25
Mining pools, 80
Mixed strategies, 22
  in extensive-form games, 92

in normal-form games, 24
Monotone hazard rate condition (MHR), 200

Moore's law, 78
Morra game, 25
Myerson's lemma, 195
  monotonic efficient allocation rule, 195

N
Nash equilibrium, 11, 33, 34, 92, 96
  computation, 37
  existence, 36
  in blockchains, 81
  in buyback contracts, 75
  in inventory games, 59
  in revenue-sharing contracts, 77
  in supply chain contracts, 71
  in wholesale-price contract, 74
  strict Nash equilibrium, 34
  weak Nash equilibrium, 34
Newsvendor games, 58
Newsvendor problem, 6, 57, 71
Nonatomic games, 164
Noncooperative games, 12
NP-Complete, 206
Nucleolus, 130

O
Operations management, 1
Optimal mixed strategy
  computation, 28
  in two-person zero-sum game, 26

P
Pareto optimality, 30, 31
Physical internet, 222
Postponement, 7
PPAD complexity class, 43
Prediction markets, 219
Preferences and utility, 13
Prisoners' dilemma, 21, 89
Process, 1
  cost, 1
  cycle time, 1
  flexibility, 1
  quality, 1
Procurement problem, 171
  direct mechanism, 188
Producer responsibility
  collective producer responsibility (CPR), 151

Index

extended producer responsibility (EPR), 150
individual producer responsibility (IPR), 151
Product-process matrix, 3
Proof-of-Work, 80
Prospect theory, 14
Public-key cryptography, 79
Pure strategies
  in extensive-form games, 91
  in normal-form games, 24

Q

Quasilinear preferences, 181

R

Rank-dependent utility theory, 14
Recycling coalitions, 150
  EcologyNet Europe, 151
  European Recycling Platform, 151
  REPIC, 151
Recycling game, 126, 150
Recycling network, 152
Repositioning games, 159
Representations of games, 14
  characteristic-form, 17
  extensive-form, 15
    of imperfect information, 16
    of perfect information, 16
  normal-form, 14
    of complete information, 48
    of incomplete information, 48
Restaurant game, 34
Revelation principle, 186
  direct mechanisms, 186
  for Bayesian-Nash equilibrium, 187
  for dominant strategy equilibrium, 187
Revenue equivalence theorem, 198
Rock-paper-scissors game, 90

S

Sage's judgment, 183
Sequential equilibrium, 99, 101
  consistency, 101
  system of beliefs, 100
Sequential rationality, 91
Service systems, 148
Shapley value, 12, 125, 133
  axioms, 135
  greedy algorithm, 134
  in Aumann-Shapley pricing, 164

in GREEN game, 155
in inventory games, 147
in logistic networks, 157
in service systems, 149
in SHAP, 160
SHAP (SHapley Additive exPlanations), 159

  axioms, 161
  feature attribution, 160
  feature importance, 160
  in quality modeling, 161
Signaling games, 107
Simple game, 125
Social choice function, 181
Sponsored search markets, 211
  click-through rate, 212
  keyword-based advertising, 211
  pay-per-click, 212
  slate, 215
Strategy functions, 48, 51
Subadditive games, 126, 148, 164
Subgame, 96
Subgame perfect Nash equilibrium (SPNE), 33, 95, 98
  in capacity decisions, 108
  in leadership in supply chains, 114
Superadditive games, 17
Supply chain, 7, 70
Supply chain contracts, 7
  buyback contract, 74
  revenue-sharing contract, 75
  wholesale-price contract, 73
Symmetric auctions, 204

T

Telephone billing game, 167
Trade game, 126
Traffic planning, 61
Transaction costs, 70
Transfer payments, 71
Transshipment, 155
Trustless consensus-building mechanisms, 78
Two-part tariff contracts, 115
Two-person zero-sum game, 22, 25, 28

U

Ultimatum game, 93

V

VCAP framework, 18

Index

Vendor-managed inventory (VMI), 8
Vertical competition, 67
Vertical integration, 7, 70
Veto player, 125
Vickrey–Clarke–Groves (VCG) mechanism, 207
  in sponsored search markets, 217
vNM expected utility theory, 14
von Stackelberg duopoly, 86, 121
  von Stackelberg follower, 86
  von Stackelberg leader, 86

Voting game, 125

W
Winner determination problem
  in combinatorial auctions, 205
  in physical internet, 226

Z
Zermelo's theorem, 92